Disclaimer

Book Title: The Process of Verification and Validation of Building Fire Evacuation Models

Book Author: Enrico Ronchi; Erica D. Kuligowski; Paul A. Reneke; Richard D. Peacock; Daniel Nilsson;

Book Abstract: To date, there is no International standard on the methods and tests to assess the verification and validation (V&V) of building fire evacuation models, i.e., model testers adopt inconsistent procedures or tests designed for other model uses. For instance, the tests presented within the MSC/Circ.1238 Guidelines for evacuation analysis for new and existing passenger ships provided by the International Maritime Organization are often employed for the V&V of models outside their original context of use (building fires instead of maritime applications). This document discusses the main issues associated with the definition of a standard procedure for the V&V of building fire evacuation models. A review of the current procedures, tests (e.g. the MSC/Circ.1238 Guidelines), and methods available in the literature to assess the V&V of building evacuation models is provided. The capabilities of building evacuation models are evaluated studying their five main core components, namely 1) Pre-evacuation time, 2) Movement and Navigation, 3) Exit usage, 4) Route availability and 5) Flow constraints. A set of tests and recommendations about the verification of building evacuation models is proposed. Suggestions on simple qualitative validation tests are provided together with examples of experimental data-sets suitable for the analysis of different core components. The uncertainties associated with evacuation modelling are discussed. In particular, a method for the analysis of behavioural uncertainty (uncertainty due to the use of distributions or stochastic variables to simulate human behaviour in evacuation modelling) is presented. The method consists of a set of convergence criteria based on functional analysis. The last part of this document presents a discussion on the definition of the acceptance criteria for a standard V&V protocol.

Citation: NIST TN - 1822

Keywords: Evacuation; Modelling; Verification; Validation; Human behaviour in fire; Building fires

NIST Technical Note 1822

The Process of Verification and Validation of Building Fire Evacuation Models

Enrico Ronchi
Erica D Kuligowski
Paul A Reneke
Richard D Peacock
Daniel Nilsson

National Institute of
Standards and Technology
U.S. Department of Commerce

NIST Technical Note 1822

The Process of Verification and Validation of Building Fire Evacuation Models

Enrico Ronchi
Daniel Nilsson
Department of Fire Safety Engineering and Systems Safety
Lund University
Lund, Sweden

Erica D Kuligowski
Paul A Reneke
Richard D Peacock
Fire Research Division
Engineering Laboratory

November 2013

U.S. Department of Commerce
Penny Pritzker, Secretary

National Institute of Standards and Technology
Patrick D. Gallagher, Under Secretary of Commerce for Standards and Technology and Director

National Institute of Standards and Technology Technical Note 1822
Natl. Inst. Stand. Technol. Tech. Note 1822, 84 pages (November 2013)
CODEN: NTNOEF

Abstract

To date, there is no International standard on procedures and tests to assess the verification and validation (V&V) of building fire evacuation models. Often it is the case that model testers adopt inconsistent procedures, or tests designed for other model uses or they do not test them for all features embedded in their model. For instance, the tests presented within the MSC/Circ.1238 (Guidelines for evacuation analysis for new and existing passenger ships) provided by the International Maritime Organization are often employed for the V&V of models outside their original context of use (e.g. building fires instead of maritime applications). This document is intended to open a discussion on the main issues associated with the definition of a standard procedure for the V&V of building fire evacuation models. A review of the current procedures, tests and methods available in the literature to assess the V&V of building evacuation models is provided. The capabilities of building evacuation models are evaluated by studying their five main core components, namely 1) pre-evacuation time, 2) movement and navigation, 3) exit usage, 4) route availability and 5) flow constraints. A set of tests and recommendations about the verification and validation of building evacuation models is proposed. These tests include suggestions on using simple tests of emergent behaviours together with examples of experimental data-sets suitable for the analysis of different core components. The uncertainties associated with evacuation modelling are discussed. In particular, a method for the analysis of *behavioural uncertainty* (uncertainty due to the use of distributions or stochastic variables to simulate human behaviour in evacuation modelling) is presented. The method consists of a set of convergence criteria based on functional analysis. The last part of this document presents a discussion on the issues associated with the definition of the acceptance criteria of a standard V&V protocol.

Keywords
Evacuation, Modelling, Verification, Validation, Human behaviour in fire, Building fires

Acknowledgement

The authors wish to acknowledge Jason Averill, Thomas Cleary, Rita Fahy, Anthony Hamins, Steve Gwynne, Blaza Toman and Craig Weinschenk for a comprehensive review of this document before publication.

Table of Contents

1.0 Introduction

The process of verification and validation (V&V) is a key factor when assessing the reliability of the results produced by simulation models and defining their suitable fields of application. Evacuation models are no exception. The meaning of *verification* is *"the process of determining that a calculation method implementation accurately represents the developer's conceptual description of the calculation method and the solution to the calculation method* [International Standards Organization, 2008]. This definition is globally accepted in the context of the fire safety engineering community and the sub-field of evacuation modelling [International Standards Organization, 2008].

The definition of validation in the context of simulation models is somewhat ambiguous [Rykiel, 1996]. Validation is defined as the *"process of determining the degree to which a calculation method is an accurate representation of the real world from the perspective of the intended uses of the calculation method"* [International Standards Organization, 2008]. Several questions can be prompted by this definition:

- How do we judge if a tool is accurate enough (i.e., the definition of the acceptance criteria)?
- How many and which tests should be performed to assess the accuracy of the model predictions?
- Who should perform these tests i.e., the model developers, the model users or a third party?

These questions do not have simple answers and the assessment of "validity" of evacuation models is a difficult task. It requires, even for a very small field of application, reliable experimental data-sets, a common method of analysis of the model predictions, and the definition of the acceptance criteria.

Whereas the process of verification can be made by employing a set of hypothetical test cases [International Maritime Organization, 2007], validation testing generally relies on the availability of experimental data and the subsequent uncertainties associated with them. Two main aspects need to be discussed about validation, i.e., the field of application (e.g., the evacuation scenarios) and the duration of validity. Lord et al., [2005] conducted work aimed at addressing the uncertainty and variability in egress data and modelling. They studied the sources and types of uncertainty in a set of evacuation models and they tested their predictive capabilities for a range of scenarios. Nevertheless, they highlighted the need for further research on the subjects stating that *"validation for one application or scenario does not imply validation for different scenarios"* [Lord et al., 2005]. The need for dedicated validation efforts in relation to different scenarios and the evolving characteristics of the models are also discussed by Galea [1997] who stresses the need for on-going validation efforts for egress models as the models and our understanding of human behaviour during evacuation advances.

Existing data for the validation of evacuation models are limited [Averill et al., 2008]. In addition, there is not a common standard procedure to perform V&V specifically designed for building evacuation models. To date, the International Standards Organization provided a document [International Standards Organization, 2008] with only general information on the assessment and V&V of calculation methods in the context of fire safety engineering.

The main guidance available that described the V&V of evacuation models is currently provided by the International Maritime Organization (IMO), namely the MSC/Circ.1238 (Guidelines for evacuation analysis for new and existing passenger ships) [International Maritime Organization, 2007]. These guidelines describe the V&V of maritime evacuation simulation tools, but they are often employed for testing evacuation models for other application areas (e.g., buildings, other means of transportation, etc.).

The MSC/Circ.1238 lists four main forms of tests that have to be performed for evacuation models, namely 1) component testing, 2) functional verification, 3) qualitative verification and 4) quantitative verification. In this paper, a different term, based on the definition included in ISO 16730, is used, i.e., *quantitative validation* [International Standards Organization, 2008]. Component testing is the process of checking that the components of a model work as intended. Functional verification involves the analysis of the capabilities of the model in terms of its ability to perform the intended simulations. Qualitative and quantitative verification (quantitative validation here) regards the nature of predicted human behaviour with informed expectations. While guidelines provide information on methods to assess the first three forms of tests, they do not provide information on quantitative verification (i.e. the comparison between model predictions against reliable experimental data). In particular, they do not investigate uncertainty in the model results produced that is associated with the possible variability of human behaviour in fire (often addressed and represented using a stochastic approach).

The main issue is that these tests are specifically designed for maritime applications, still required the development of V&V tests specific to building evacuations. Initial attempts have been made to improve the MSC/Circ.1238 and extend their use to different contexts of application, i.e., not only maritime applications. First, the RIMEA project [Meyer-Koenig et al., 2007] has reviewed the MSC/Circ.1238 and has proposed modifications in the tests to be performed. However, the modifications proposed within the RIMEA project do not include validation tests. In addition, the revised tests do not include new tests on many of the features that are generally available in building evacuation models. Second, project SAFEGUARD [Galea et al., 2012a, Galea et al., 2012b] has been carried out in order to review the

2

MSC/Circ.1238 and provide suggestions on possible improvements. However, the improvements still focus on maritime applications and the acceptance criteria may not be suitable for every model or model scenario.

The consequence is that, given the lack of an International Standards on the tests and methods to assess the V&V of building evacuation models, model developers currently adopt inconsistent procedures. Therefore, efforts are made here to design a standard procedure for the verification and validation of evacuation models used for building evacuation. V&V model documentation based on a standard procedure would assist model users in model selection. In this paper, suggestions are provided on the methods and tests that can be employed to assess model capabilities. The definition of standard V&V tests and procedures requires a broad debate among all parties involved in the evacuation community (e.g., evacuation modellers, experimentalists, building owners, authorities having jurisdiction, etc.). For this reason, this document is aimed at opening a discussion on the topic rather than being a definitive guidance on the topic.

1.1 Objectives

This document suggests a standard procedure for the verification and validation of building evacuation models. The overall scope is not to provide definitive guidance on the performance of V&V for building evacuation models. The present work is instead designed to open a debate on the issues associated with V&V for building evacuation models. In fact, there is a need to develop a validation protocol and acceptance criteria which are accepted by all parties involved, namely model users, model developers, regulators, etc. In this context, this document has also been prepared to contribute to the on-going effort made by the ISO/TC92/SC4 Working Group 7 on the further development of the ISO 16730 document "Fire Safety Engineering - Assessment, verification and validation of calculation methods" [International Standards Organization, 2008].

The first objective of this document is to review the main procedures, tests, and methods available in the literature to assess the V&V of building evacuation models (e.g. the tests presented within the MSC/Circ.1238 [International Maritime Organization, 2007].

The second objective is to present a new set of tests for the verification of building evacuation models and provide examples of data-sets and methods for validation studies. Some of the verification tests are based on the tests provided within the MSC/Circ.1238 [International Maritime Organization, 2007]. The capabilities of building evacuation models are studied by analysing their five main core components [Gwynne et al., 2012a], namely 1) pre-evacuation time, 2) travel speed, 3) exit usage, 4) route availability and 5) flow conditions/constraints.

The third objective is to define a method for the study of a type of uncertainty which is typical of evacuation modelling (associated with the use of distributions or stochastic variables to simulate human behaviour), here named *behavioural uncertainty*. The term *behavioural uncertainty* is used to highlight the uncertainty associated with the current lack of knowledge in evacuation research about some of the variables affecting human behaviour of individuals and populations [Ronchi et al., 2013c]. In fact, to date, evacuation experimental research does not permit in all cases a full prediction of the variance associated with human behaviour. The use of the term *uncertainty* (rather than behavioural variance or behavioural variability) has been chosen since this document refers to evacuation simulations, intended as a tool to characterize the evacuation behaviour given that some aspects of the system are not exactly known. A method for the study of behavioural uncertainty in evacuation model results is necessary to define a validation protocol.

1.2 Limitations

The definition of a standard procedure for the verification and validation of building evacuation models faces a number of challenges. The first issue is that evacuation modelling is a relatively new field of science and the capabilities of evacuation models are rapidly developing [Kuligowski et al., 2010]. This is reflected in a continuous development of the model features. There is, therefore, a difficulty in developing a comprehensive list of tests which are able to evaluate their evolving capabilities.

While the definition of an exhaustive list of verification tests for building evacuation models is theoretically possible (tests might be defined for each evacuation model in order to evaluate every single embedded sub-algorithm), a number of different issues reduce the likelihood of doing so in practice. To date, behavioural experimental data-sets are scarce, i.e., data on human behaviour in building fires are limited in quantity and quality [Averill et al., 2008]. This contributes to the current lack of a *"robust, comprehensive and validated conceptual model of occupant behaviour during building fires"* [Kuligowski, 2013]. This problem impacts the capabilities of evacuation models which may be based on user-defined assumptions rather than providing a prediction of human behaviour in fire.

Another issue is associated with the definition of the acceptance criteria in relation to the context of use. In fact, different evacuation model use may require different acceptance criteria. For instance, evacuation models can be used for forensic analyses (reconstruction of an actual fire evacuation) or the calculation of the Required Safe Egress Time (RSET), i.e., the time needed by building occupants to reach a safe place [Gwynne et al., 2012b].

The present document should not be considered as definitive guidance on the procedure to perform the verification and validation of building evacuation models. Its aim is instead to discuss the issues associated with the definition of a V&V protocol for building evacuation models and provide initial assistance to model developers, users and regulators on the methods and tests that can be employed given the current state-of-the-art of fire evacuation research.

1.3 Outline

The present technical note is divided in five sections. Already presented in Section 1 (Introduction) was an introduction to the research problem, identifying the need for a standard procedure for the V&V of building evacuation models, and challenges associated with this process.

In Section 2 (Background and Previous Research), relevant literature is discussed. This includes a review of the tests provided within the MSC/Circ.1238 [International Maritime Organization, 2007], which is often considered the reference for the V&V of evacuation models. Current methods adopted for the analysis of evacuation model results are also reviewed.

Section 3 (Suggested Verification and Validation Tests) presents a new set of recommended verification tests and discusses possible examples of validation tests. Tests have been presented in relation to the five main core elements available in evacuation models, namely 1) pre-evacuation time, 2) movement and navigation, 3) exit usage, 4) route availability and 5) flow conditions/constraints.

In Section 4 (Uncertainty in Evacuation Modelling), the issues associated with uncertainty of experimental data and modelling results in the context of evacuation research are analysed. In particular, the uncertainty derived from simulating human behaviour (*behavioural uncertainty*) is discussed and method for the study of behavioural uncertainty in evacuation modelling is presented.

Section 5 (Discussion on the V&V protocol) reviews the current literature about the verification and validation protocols employed in evacuation modelling. This section also provides an analysis of the issues associated with the definition of acceptance criteria.

2.0 Background and Previous Research

This section presents a review of the current tests adopted for the analysis of the predictive capabilities of evacuation models. In particular, this section reviews the tests provided within the MSC/Circ.1238 [International Maritime Organization, 2007], which are currently considered the benchmarks available for the assessment of evacuation model capabilities.

In the present work, the analysis of the main elements and uses of building evacuation models [Kuligowski et al., 2010, Ronchi and Kinsey, 2011] was used to identify the tests required. Tests are divided into two main groups: 1) Verification tests and 2) Validation tests.

In order to perform a systematic review of the tests currently adopted for the verification and validation of building evacuation models, the core performance behavioural elements that are part of egress analysis needs to be defined. Gwynne et al. [2012a] have identified five main elements, namely, 1) pre-evacuation time, 2) travel speed, 3) exit usage, 4) route availability and selection, and 5) flow conditions/constraints. These elements were chosen as topics for consideration for V&V in this paper to represent the aspects that can be addressed in the majority of egress models (from hydraulic calculations [Gwynne and Rosenbaum, 2008] to simulation tools [Kuligowski et al., 2010]). A detailed assessment of building fire evacuation model capabilities needs to consider travel speeds under different conditions (e.g., different level of congestions, smoke, etc.). For this reason, the second element *Travel speed* has been renamed here as a more general *Movement and Navigation*.

The five core behavioural performance elements may be compared against ideal cases or experimental data when performing V&V. Ideal (i.e., hypothetical) cases are defined here as simple evacuation scenarios for which the expected result can be derived by simple mathematical formulae or evidence derived from the current knowledge on human behaviour in fire. Assuming that the underlying element being verified is simple, they can be used to assess model capabilities. For example, an ideal case can be the time employed by a single agent to walk a corridor at a certain walking speed. Ideal cases can also be employed to perform the study of emergent behaviours, i.e. the evaluation of a sub-algorithm embedded in a model. For example, an ideal test case can be used to qualitatively analyse the impact of group behaviours. The use of ideal test cases for the study of emergent behaviours within building fire evacuation models may be driven by the current lack of experimental data[1] suitable for the comparison with model predictions. In those cases, ideal cases may be employed to perform a qualitative evaluation of the predictive capabilities of the models.

---

[1] In the present document, the term experimental data (including evacuation drills) is used since real evacuation data are scarce and they are rarely employed for validation purposes. The term "real evacuation data" is used when referring to actual evacuation scenarios (data from evacuation drills are not included in this category).

Ideally, the quantitative validation of the main behavioural components should always be performed using experimental (or real evacuation) data. Evacuation experimental data include data-sets derived from several types of data such as laboratory experiments (i.e., experiments performed in a controlled laboratory), field experiments (i.e., experiments performed in real-life settings), or real evacuation data [Nilsson, 2009]. The degree of control on the experiments together with the data collection methods may significantly affect the uncertainty in the results. The suitability of experimental data for the performance of validation tests relies on uncertainty of the data (e.g., based on the background conditions, the data collection techniques adopted, etc.), the documentation provided with the experimental data-sets and their availability to the public. The comparison between simulation results and experimental data should, therefore, consider the limitations associated with the experiments.

Another aspect affecting the study of simulation results are the factors associated with human behaviour. Current methods (e.g. the MSC/Circ.1238 [International Maritime Organization, 2007]) do not fully investigate the impact of the use of distributions/stochastic variables employed by evacuation models to represent human behaviour [Averill, 2011]. In fact, a single experiment or model run may not be representative of a full range of the occupant behaviours. Therefore, there is a need to develop a method to perform a comparison between simulation results and experimental data which takes this into account.

In V&V, considerations should also be made on the modelling methods used by the evacuation models to produce evacuation results. For example, additional requirements/tests may be needed in relation to the type of grid/structure employed by the evacuation model. Evacuation models can be categorised in accordance with three main types of grid/structures [Kuligowski et al., 2010]: coarse, fine and continuous grids. Coarse grids use an abstract network of nodes and arcs to represent the space. Fine grids represent the space as a grid of cells. Continuous grids employ a system of coordinates to represent the space. Recent studies [Chooramun, 2011] have also implemented a new type of grid/structure, namely 4) hybrid models, i.e. two or more of the previous grids/structures are employed simultaneously in a single evacuation model.

The assumed grid/structure of a model may significantly affect the results for some specific core elements (e.g. especially the representation of 2) movement and navigation and 5) flow conditions/constraints [Lord et al., 2005]). For this reason, specific recommendations on the methods to perform V&V tests should be made to account for the intrinsic differences among the models.

2.1   Current State of V&V Tests: The IMO tests

The test cases presented in the MSC/Circ.1238 [International Maritime Organization, 2007] are reviewed and discussed in this section with the aim of evaluating their suitability and suggesting improvements for the validation and verification of building evacuation models (i.e., outside the maritime context). The MSC/Circ.1238 includes seven tests (IMO Tests 1-7) on component testing and four tests on qualitative verification (IMO Tests 8-11). Recommendations on the methods to provide functional verification are also provided. In contrast, the MSC/Circ.1238 claims that, there is *"insufficient reliable experimental data to allow a thorough quantitative verification of egress models"*. The SAFEGUARD project [Galea et al., 2012a, Galea et al., 2012b] has recently been carried out to recommend a set of experimental data to be added in the MSC/Circ.1238 and address quantitative validation. Nevertheless, those data-sets refer only to the maritime context, thus may not apply to the building context.

Initially, this document reviews the IMO tests and suggests possible improvements for their application to buildings. The values presented in each test are also evaluated in relation to the range of possible experimental values available in the evacuation literature for each corresponding variable.

IMO Test 1 is a verification test designed for testing unimpeded walking speeds:
"One person in a corridor 2 m wide and 40 m long with a walking speed of 1 m/s should be demonstrated to cover this distance in 40 s."

This test is useful to verify if the model is able to represent an agent maintaining an assigned speed over time since it is a critical aspect during the calculation of the Required Safe Egress Time of a building [Gwynne et al., 2012b]. The test presents a value of walking speed that may be representative of the walking speed of an adult (1 m/s) and a length of the corridor sufficient to test if the assigned agent speed is kept over time. This test is designed to analyse the results of a deterministic input (i.e. a single model run). The use of distributions of walking speeds may need the analysis of the results of multiple runs in the case of stochastic models.

The effectiveness of this test can be improved by setting additional prescriptions in relation to the modelling method employed. For example - in the case of coarse and fine network models results may be dependent on the configuration of the grid adopted, e.g. the relative rotation of a corridor in relation to the grid in use [Ronchi et al., 2013b]. Hence, the test of the assigned speed should be completed in conjunction with the representation of the geometry. The test should be, therefore, performed using different relative rotations of the geometry, if possible. Considerations should also be made on the necessity (or not) of performing this test with different grid configurations (e.g. the default cell size and a reduced/increased cell size) in order to test the sensitivity of the results to the cell size. Lord et al. [2005] have shown that the results of a fine network model may be dependent on the type of grid employed.

IMO Test 2 and IMO Test 3 are verification tests that examine the ability of the model to represent agent movement on a staircase:

"One person on a stair 2 m wide and a length of 10 m measured along the incline with a walking speed of 1 m/s should be demonstrated to cover this distance in 10 s."

These verification tests are useful to test specified travel speeds for people moving up and down stairs. The value of walking speed presented in the test is in the upper boundary of possible experimental walking speeds on stairs [Peacock et al, 2012]. These tests can be extended by adding the requirement to test unconventional stair designs which may be available in buildings (e.g. spiral stairs, curved stairs, etc.). Also in this case, the underlying grid employed needs to be considered, i.e. is there a need to assess the impact of different grids on the results produced including the representation and rotation of the geometry, type of the grid, cell size, etc.

IMO Test 4 is a verification test aimed at analysing a single exit flow rate:
"100 persons (p) in a room of size 8 m by 5 m with a 1 m exit located centrally on the 5 m wall. The flow rate over the entire period should not exceed 1.33 p/s."

IMO Test 4 is a test to verify a simple door flow problem. The admitted maximum flow rate in the test is in line with current evacuation literature [Gwynne and Rosenbaum, 2008]. Flow rates are modelled in evacuation models in different ways (typically either based on restricted maximum flows or being generated by the modelling assumptions employed). For this reason, this test should be considered part of component testing only if the method to represent flow rates employed by the model under consideration is based on restricted flows. If flows are emergent, the assignment of a maximum flow rate can be intended as the setting of an external conservative requirement (i.e. in the case of emergent flows without a restriction). This test may also be influenced by the type of underlying grid employed. There is a need to evaluate if there are alternative tests which may be suitable for the analysis of the flow problem. For example, a different test about flow rate is suggested within the RIMEA project [Meyer Koenig et al., 2007] based on the analysis of the comparison of simulation results of any type of model with fundamental diagrams [Weidmann, 1993].

IMO Test 5 is a verification test for the analysis of pre-evacuation times:
"Ten persons in a room of size 8 m by 5 m with a 1 m exit located centrally on the 5 m wall. Impose response times as follows uniformly distributed in the range between 10 s and 100 s. Verify that each occupant starts moving at the appropriate time."

This is a useful test to verify the ability of evacuation models to reproduce imposed pre-evacuation times. The choice of the pre-evacuation time range is reasonable since the scope of this test is to verify distribution assignment. A possible improvement to the test involves the consideration of distributions. Experimental data [Purser, 2001] shows in fact that occupant pre-evacuation times in building can be generally represented using log-normal or normal distributions. This is reflected in evacuation models which often adopt these types of distributions.

IMO Test 6 is a test to verify that occupants successfully navigate around a corner, i.e. it is a test of the boundaries of the simulated scenario:

"Twenty persons approaching a left-hand corner will successfully navigate around the corner without penetrating the boundaries."

This test is designed to verify whether the model is able to correctly simulate the boundaries of a scenario; i.e., that evacuees do not artificially cross boundaries when turning, especially in a crowded environment. The current form of the test does not add any requirement on the expected pattern of the agents, i.e. the usage of the area of the corner in relation to the number of agents in the scenario. In addition to the appropriate simulation of the constraints provided by the boundary upon agent movement, experimental studies have shown that occupants may occupy only a portion of the space available in the corner in relation to different conditions, e.g. the space usage can be affected by the observed densities [Nilsson and Petersson, 2008, Zhang et al., 2011]. An additional test or a modified version of the current test could be defined in order to investigate this issue.

IMO Test 7 is aimed at verifying the correct assignment of population demographics parameters:

"Choose a group consisting of males 30-50 years old from table 3.4 in the appendix to the IMO Guidelines [International Maritime Organization, 2007] for the advanced evacuation analysis of new and existing ships and distribute the walking speeds over a population of 50 people. Show that the distributed walking speeds are consistent with the distribution specified in the table."

This test aims to verify that the characteristics of occupants are consistent with the assigned values. In order to apply this test in the building context, there is a need to re-evaluate the population type or characteristics employed in relation to the different type of environment (buildings) under consideration.

IMO Test 8 is a verification test about counter-flows:

"Two rooms 10 m wide and long connected via a corridor 10 m long and 2 m wide starting and ending at the centre of one side of each room (see International Maritime Organization, [2007] for further information on the geometry). Choose a group consisting of males 30-50 years old from table 3.4 in the appendix to the Guidelines [International Maritime Organization, 2007] for the advanced evacuation analysis of new and existing ships with instant response time and distribute the walking speeds over a population of 100 persons. Step 1: One hundred persons move from room 1 to room 2, where the initial distribution is such that the space of room 1 is filled from the left with maximum possible density. The time the last person enters room 2 is recorded. Step 2: Step one is repeated with an additional ten, fifty, and one hundred persons in

room 2. These persons should have identical characteristics to those in room 1. Both rooms move off simultaneously and the time for the last persons in room 1 to enter room 2 is recorded. The expected result is that the recorded time increases with the number of persons in counter-flow increases."

This test is useful to qualitatively verify the ability of models to simulate counter-flow and its possible impact on evacuation time. Also in this case, in order to apply this test in the building context, there is a need to re-evaluate the population type or characteristics in relation to the different type of environment (buildings) under consideration. Further tests are needed to include the analysis of the merging ratios, the flow patterns produced, the extent of the increase in congestion, etc. In order to provide a complete assessment of the capabilities of building evacuation models, an additional test should be designed to test counter-flow on stairs.

IMO Test 9 consists of a test of crowd dissipation from a large public room:
"Public room with four exits and 1,000 persons uniformly distributed in the room (see International Maritime Organization, [2007] for further information on the geometry). Persons leave via the nearest exits. Choose a group consisting of males 30-50 years old from table 3.4 in the appendix to the Guidelines [International Maritime Organization, 2007] for the advanced evacuation analysis of new and existing ships with instant response time and distribute the walking speeds over a population of 1,000 persons. Step 1: Record the time the last person leaves the room. Step 2: Close doors 1 and 2 and repeat step 1. The expected result is an approximate doubling of the time to empty the room."

IMO Test 9 is a verification test to qualitatively evaluate the ability of the models to simulate the impact of a reduction in the available exits on the simulation results. This test is useful to calculate the Required Safe Egress Time in the case of a different number of evacuation designs. The characteristics of the population should be re-evaluated in order to consider building occupants. This test does not evaluate the predictive capabilities of evacuation models in terms of exit usage or the ability to assign the agents to specified exits. For instance, this test could be modified in order to represent the possible loss of the main exit during fire safety engineering analysis.

IMO Test 10 is an exit route allocation test:
"Construct a cabin corridor section (see International Maritime Organization, [2007] for further information on the geometry) populated as indicated with a group consisting of males 30-50 years old from table 3.4 in the appendix to the Guidelines [International Maritime Organization, 2007] for the advanced evacuation analysis of new and existing ships with instant response time and distribute the walking speeds over a population of 23 persons. The people in cabins 1, 2, 3, 4, 7, 8, 9, and 10 are allocated the main exit. All the remaining passengers are allocated the secondary exit. The expected result is that the allocated passengers move to the appropriate exits."

IMO Test 10 is aimed at verifying the ability of the model to reproduce exit choice in a deterministic way. For instance, this test is useful for the representation of exit usage given the loss of the main exit during fire safety engineering analysis. Also in this case, the population type and characteristics should be re-evaluated in order to be representative of buildings. Additional tests on exit allocation can be designed in order to consider other factors that may impact route usage/exit choice, such as the presence of smoke [Ronchi et al., 2013a], the presence of way-finding installations [Nilsson, 2009], social influence [Nilsson and Johansson, 2009], affiliation [Sime, 1984], etc.

IMO Test 11 is a test aimed at verifying the flow constrains in a staircase:
"Construct a room connected to a stair via a corridor (see International Maritime Organization, [2007] for further information on the geometry) populated as indicated with a group consisting of males 30-50 years old from table 3.4 in the appendix to the Guidelines [International Maritime Organization, 2007] for the advanced evacuation analysis of new and existing ships with instant response time and distribute the walking speeds over a population of 150 persons. The expected result is that congestion appears at the exit from the room, which produces a steady flow in the corridor with the formation of congestion at the base of the stairs."

This test is useful to verify the capabilities of evacuation models in terms of reproducing congestion. One limitation is that the test is only qualitative and it does not quantify the form and size of the congestion. The population should be modified in order to be representative of building evacuations. A test including movement upstairs may not be representative of the expected scenarios in a building fire evacuation study, where occupants generally move downstairs. Therefore, this test can be modified in order to include movement downstairs.

The present analysis of the IMO tests discusses their applicability in the building context. The analysis revealed that IMO Tests can be a starting point for the assessment of the capabilities of building fire evacuation models. Nevertheless, they do not cover all features included in evacuation models for building applications and they are designed to test populations which are specific to the maritime context. This is in contrast with their current use for the verification of evacuation models in the building context. For this reason, the following section presents a set of new tests which are mostly based on the IMO Tests but they present modifications in order to extend their application to buildings.

3.0 Suggested Verification and Validation Tests

The previous section presented a review of the tests suggested in the MSC/Circ.1238 [International Maritime Organization, 2007] in order to highlight and discuss their limitations. The current section expands and modifies the list of tests presented in order to extend their applicability to building evacuations. A set of suggested verification tests (Section 3.1) is provided. Following that, a set of examples of possible validation tests (Section 3.2) are described.

The verification tests are either based on the tests presented in the MSC/Circ.1238 [International Maritime Organization, 2007] or they are newly-developed tests by the authors of this report. The selection of the additional tests is made in order to include the features listed in the review of building evacuation models by Kuligowski et al. [2010]. The review includes the following features: 1) Counterflow, 2) Exit block, 3) Fire conditions affecting behaviour, 4) Toxicity, 5) Groups, 6) Disabled/slow occupants, 7) Delays/Pre-evacuation times, 8) Elevator use, 9) Route choice. Additional tests are considered if critical factors which may substantially affect RSET are identified.

The tests are structured in five parts: 1) *Geometry*: the configuration of the test, 2) *Scenario(s)*: the evacuation scenario that is going to be simulated, 3) *Expected result*: the result (qualitative or quantitative) that the evacuation model is supposed to produce, and 4) *Test method*: the qualitative (e.g., visualization of the represented behaviour) or quantitative (e.g., comparison of evacuation times, flows, etc.) method employed for the comparison between the expected result and the simulation results, 5) *User' actions*: the actions required of the tester while performing and presenting the tests.

It should also be noted that different models may require different test methods for the analysis of their results regarding the input variables required for their calibration. Calibration is here intended as the set of actions required to model developers/users to configure the model prior to its use. The model tester is required to list the user's actions during the calibration of the input expected while running the tests. For example, models employing a deterministic user-defined representation of expected behaviours (e.g. the agents move towards a user-assigned exit) can be tested performing quantitative verification tests only, i.e. no validation studies can be made. On the other hand, models including predictive sub-models can be tested using both qualitative and quantitative validation testing.

3.1   Verification Tests

This section presents the tests suggested for the verification of evacuation models. The tests are organized using the five main core components of evacuation models [Gwynne et al., 2012a], namely 1) pre-evacuation time, 2) movement and navigation, 3) exit usage, 4) route availability and 5) flow conditions/constraints; i.e. they are the elements of a model required for the most basic representation of a scenario. They are ideal (hypothetical) tests that are designed to analyse the main features of current evacuation models. Some of the tests are based on the MSC/Circ.1238 [International Maritime Organization, 2007]. The tests are divided into two categories. The first category is called analytical verification (AN_VERIF) and it refers to component testing where the expected results can be derived by simple mathematical formulae or evidence. The second category is the verification of emergent behaviours (EB_VERIF), which refers to the verification of the ability of evacuation models to qualitatively produce results which reflect the current knowledge on human behaviour in fire. In the present work, this second category of tests is deliberately not labelled as validation. This is because those tests evaluate if model results are in line with current behavioural theories rather than making a direct quantitative use of experimental (or real) evacuation data.

It should be noted that not all of the tests can be conducted for all of the models available given the different functionality available within the models. The term *occupant* is used to refer to a general model agent whose main physical characteristics are walking speeds and body size. Evacuation models generally characterize gender as a consequence of the assumed body size and walking speeds, not in terms of possible behavioural differences. For this reason, the gender of the occupants is not explicitly mentioned in the tests.

Table 1 presents the suggested tests in relation to the core components and the sub-elements under consideration.

Table 1. Suggested verification tests for evacuation models.

Core component	Sub-element	Suggested tests	Test code	Type of Test
1	Pre-evacuation time distributions	Modified IMO Test 5	Verif.1.1	AN_ VERIF
2	Speed in a corridor	IMO Test 1	Verif.2.1	AN_ VERIF
	Speed on Stairs	IMO Test 2 and IMO Test 3 (if necessary)	Verif.2.2	AN_ VERIF
	Movement around a corner	IMO Test 6	Verif.2.3	AN_ VERIF
	Assigned demographics	Modified IMO Test 7	Verif.2.4	AN_ VERIF
	Reduced visibility vs walking speed	New test	Verif.2.5	AN_ VERIF
	Occupant incapacitation	New test	Verif.2.6	AN_ VERIF
	Elevator usage	New test	Verif.2.7	AN_ VERIF
	Horizontal counter-flows (rooms)	Modified IMO Test 8	Verif.2.8	EB_ VERIF
	Group behaviours	New test	Verif.2.9	EB_ VERIF
	People with movement disabilities	New test	Verif.2.10	EB_ VERIF
3	Exit route allocation	Modified IMO Test 10	Verif.3.1	AN_ VERIF
	Social influence	New test	Verif.3.2	EB_ VERIF
	Affiliation	New test	Verif.3.3	EB_ VERIF
4	Dynamic availability of exit	New test	Verif.4.1	AN_ VERIF
5	Congestion	Modified IMO Test 11	Verif.5.1	EB_ VERIF
	Maximum flow rates	IMO Test 4	Verif.5.2	EB_ VERIF

The suggested test methods employed for the verification of evacuation models consists of:

1) A quantitative evaluation of model results. This evaluation is generally expressed as the percentage of differences between the expected results and the simulation results.

2) A qualitative evaluation of model results. This test method relies on qualitative observations of an expected behaviour. This evaluation is assessed based upon a comparison of the evacuation results (often performed via observation of the model's

visualization output or numerical results) and expected behaviours based upon the current behavioural theory.

It is recommended that the results of the tests are associated with a detailed documentation of the modelling assumptions employed to perform the tests. In particular, a description of the characteristics of the sub-algorithm(s)/sub-model(s) in use would permit understanding the model capabilities and its limitations (e.g., if and how the model performs a certain test).

3.1.1 Pre-Evacuation time

Pre-evacuation time is the time needed by the evacuees to start the movement towards a place of safety [Gwynne et al., 2012a]. One simple test (Verif.1.1.) is suggested to verify the ability of evacuation models to assign distributions of pre-evacuation times to occupants.

Verif.1.1. Pre-evacuation time distributions

This test deals with the representation of pre-evacuation times within evacuation models. The proposed test is a modified version of the IMO Test 5 from the MSC/Circ.1238.

Geometry
A room of size 8 m by 5 m with a 1 m exit.

Scenario
Ten persons are randomly located in the room. Check the types of distributions used by the evacuation model to represent pre-evacuation times. Impose a pre-defined distribution (e.g. uniform, normal, log-normal, etc.) of pre-evacuation times in accordance with the input distributions provided within the evacuation model. Repeat the test for each distribution of pre-evacuation time embedded in the model.

Expected result
Verify that each occupant starts moving at the appropriate time and that the responses of the population fall within the specified range.

Test method
The test method is a quantitative verification of the model assignment expressed in terms of pre-evacuation time. In relation to the type of distribution under consideration, the model tester needs to identify a suitable quantitative method to evaluate the differences among the simulated and assigned distributions.

<u>User's actions</u>

It should be noted that this test should be repeated several times (i.e. multiple runs of the same scenario should be done) in order to verify the simulation of the expected pre-evacuation time distributions over multiple runs[2].

3.1.2 Movement and Navigation

A total of ten tests are suggested for the verification of this core element. Seven tests are aimed at analytical verification of the models (AN_VERIF), and three tests verify the representation of emergent behaviours (EB_VERIF). The first two tests address the simulation of assigned walking speeds in a corridor (Verif.2.1) and up or down a staircase (Verif.2.2). Verif.2.3 is a test about occupant navigation around a corner. A test (Verif.2.4) is suggested for the assignment of occupant demographics. Verif.2.5 investigates the simulation of horizontal counter-flows. New tests are suggested for the verification of the impact of smoke on occupant walking speed (Verif.2.6) and the simulation of incapacitation (Verif.2.7). A verification test (Verif.2.8) of this core behavioural component deals with the simulation of elevators. The last two tests deals with the analysis of emergent behaviours of groups (Verif.2.9) and people with movement disabilities (Verif.2.10). These tests were selected in order to include the majority of movement and navigation features listed in the review of building evacuation models by Kuligowski et al. [2010].

Verif.2.1. Speed in a corridor

A test is proposed to verify the simulation of an occupant maintaining an assigned walking speed over time. The test is based on IMO Test 1 from the IMO Guidelines.

<u>Geometry</u>
A corridor 2 m wide and 40 m long.

<u>Scenario</u>
One occupant with an assigned walking speed of 1 m/s walking along the corridor.

<u>Expected result</u>
The occupant should cover the distance of the corridor in 40 s.

<u>Test method</u>

---

[2] It should be noted that the requirements for the number of runs may vary over different distributions. Different tests may be employed to demonstrate that the values belong to certain distributions. Model testers can demonstrate this with a suitable test in relation to the distribution(s) under consideration.

The test method is a quantitative verification of model results, i.e., the difference between the expected result and the simulation results.

User's actions

The effectiveness of this test can be improved by setting additional prescriptions in relation to the type of model under consideration. For example, in the case of models that use coarse and fine grids, results may be dependent on the configuration of the grid adopted. In the case of models using a fine grid, results may be affected by the rotation of the corridor in relation to the grid in use [Ronchi et al., 2013b]. The test should, therefore, be performed using at least two different rotations of the geometry (e.g., 0 degrees and 45 degrees). Considerations should also be made on the necessity (or not) to perform this test with different grid configurations (e.g. simulating the default cell size and a set of both reduced and increased cell sizes) in order to test the sensitivity of the results to cell size.

Verif.2.2. Speed on Stairs

A set of verification tests is necessary to test people movement up and down stairs. The proposed test(s) is(are) based on IMO Guidelines. IMO Test 2 and IMO Test 3 are two verification tests about maintaining a set walking speed up or down a staircase.

Geometry

A stair 2 m wide and with a length of 100 m measured along the incline.

Scenario

One occupant with a walking speed of 1 m/s (upwards or downwards) is walking along the stair.

Expected result

The occupant is expected to cover the distance in 100 s (upwards or downwards).

Test method

The test method is a quantitative verification of model results, i.e., the difference between the expected result and the simulation results.

User actions

IMO Test 2 and IMO Test 3 examine the same component. Evacuation models may use the same input to modify people movement in stairs (either upward or downward movement). For example, the user defines a speed factor (either manually or by inserting certain parameters such as stair tread and width). It could be possible to perform only one of those two tests if the models are using the same basic function to simulate the movement upwards and downwards (i.e. two tests may become unnecessary if the input employed by the model is the same). The requirement to test unconventional stair designs can be added in order to extend the applicability of building evacuation models to those scenarios (e.g. spiral stairs, curved stairs, etc.). It should also be noted that current models do not generally permit a direct representation of the impact of fatigue on walking speeds on stairs. Once this feature is implemented in the models, a corresponding verification test would need to be developed. Also in this test, the tester has to show, in the case of network models (coarse or fine network), the sensitivity of model results to the network employed and assess if the rotation of the geometry may have an impact on results.

Verif.2.3. Movement around a corner

One test is proposed to verify whether the model is able to correctly simulate the boundaries of a scenario. IMO Test 6 is the benchmark for this test, i.e., a test to verify that occupants successfully navigate around a corner.

Geometry
A corner is represented in accordance with Figure 1.

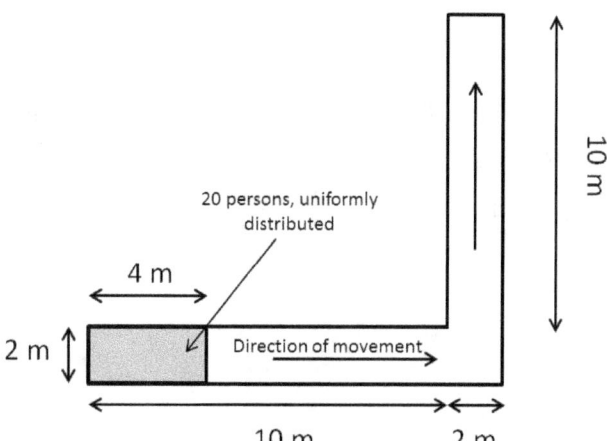

Figure 1. Geometric layout of Verif.2.3 Test. The test is the same as IMO Test 6 [International Maritime Organization, 2007].

Scenario
Twenty persons are uniformly distributed in one end of the hallway (in a space measured 2 m by 4 m). They have immediate response times and a walking speed of 1 m/s.

Expected result
The occupants are expected to successfully navigate around the corner without penetrating the boundaries.

Test method

The test method is a qualitative verification of the occupant movement. The qualitative analysis is performed by observing the travel path walked by the occupants. If possible, this evaluation can be performed using the visualization tool of the model or tracking the coordinates of the paths of the agents.

User actions

It should be noted that the current test of movement around a corner is intended only as a verification of the boundaries available in the scenario, i.e. no evaluation of the expected pattern in the corner is made (i.e. the current test is not a verification of emergent behaviours). When the literature on human behaviour in fire is able to provide a detailed understanding of the expected movement patterns of people, model testers will need to include this in the test.

Verif.2.4. Assigned occupant demographics

A test is proposed to verify the ability of the model to assign population demographic parameters. The proposed test is a modified version of the IMO Test 7.

Geometry
A squared room of size 100 m by 100 m.

Scenario
Choose a sub-population consisting of a population selected in accordance with the expected characteristics of the building(s) (see Lord et al. [2005] for possible occupant demographics). Assign the walking speeds over a population of 100 occupants evenly distributed in the room.

Expected result
Show that the assigned walking speeds are consistent with the distribution specified in the scenario.

Test method
The test method is a quantitative verification of model assignments, i.e. the analysis of the walking speeds simulated by the evacuation model. In relation to the type of distribution under consideration, the model tester needs to identify a suitable quantitative method to evaluate the differences among the simulated and assigned distributions.

User actions
It should be noted that values to be used for the characterization of occupant demographics are dependent on several factors, such as building use, nationality, etc. Please refer to Lord et al. [2005] for examples of actual distributions. Also in this case, model testers should demonstrate

that the simulation of occupant demographic distributions is verified over multiple runs, i.e., the test should be repeated several times.

Verif.2.5. Reduced visibility vs walking speed

This test is aimed at quantitatively verifying the ability of evacuation models to reproduce the physical impact of smoke upon occupant walking speeds. It should be noted that smoke has numerous additional physical, psychological and sociological factors [Ronchi et al., 2013a] that are not currently captured by evacuation models. This test is based on the reduced visibility vs. walking speed verification test suggested by Korhonen and Hostikka [2009].

Evacuation models may use different data-sets and correlations to represent the impact of reduced visibility (smoke) on walking speeds. For this reason, the tester needs to be aware of the type of correlation employed by the model in order to perform this verification test. Five types of correlations have been identified by Ronchi et al. [2013a] in current evacuation models that take into consideration different assumptions about the impact of reduced visibility on individual walking speeds and the minimum walking speed in smoke. These correlations are presented here:

$$v_i^s = v_i^0 c(K_s)$$ 　　　　　　　　　　　[Equation 1, Ronchi et al., 2013a]

$$v_i^s = Max\{v_{i,min}, v_i^0 c(K_s)\}$$ 　　　　　[Equation 2, Ronchi et al., 2013a]

$$v_i^s = Max\{v_{i,min}(i), v_i^0 c(K_s)\}$$ 　　　　[Equation 3, Ronchi et al., 2013a]

$$v_i^s = Max\{v_{i,min}, v_i(K_s) \pm \Delta\}$$ 　　　[Equation 4, Ronchi et al., 2013a]

$$v_i^s = Max\{v_{i,min}(i), v_i(K_s) \pm \Delta\}$$ 　[Equation 5, Ronchi et al., 2013a]

Where:
v_i^s is the walking speed in smoke
v_i^0 is the walking speed in clear conditions
K_s is the extinction coefficient, which relates the intensity of monochromatic light and the intensity of the light transmitted through the path-length of the smoke [Mulholland, 2008]
$c(K_s)$ is a speed reduction function (i.e. $0<c\leq1$) depending on the extinction coefficient K_s
$v_{i,min}$ is a minimum speed in dense smoke for all individuals
$v_{i,min}(i)$ is the individual minimum speed in smoke
Δ is a range of speeds around the speed under consideration

Equation 1 represents a fractional impact of smoke on speed without a minimum speed in dense smoke. n smoke/speed curves are produced in accordance with the characteristics of n individuals under consideration. Equation 2 represents a fractional impact of smoke on speed with a minimum constant walking speed in dense smoke ($\approx$0.3 m/s to 0.4 m/s); n smoke/speed curves are produced, but they present all the same minimum speed. Equation 3 represents a fractional impact of smoke on speed with a variable minimum speed in dense smoke; n smoke/speed curves are produced in accordance with the characteristics of n individuals, and the minimum speed is dependent on the characteristics of the individuals. Equation 4 represents an absolute reduction of speed in relation to the smoke, within a certain range, Δ, of speeds around the average, i.e. speed reduction is independent from the occupant speed in clear conditions. Equation 5 is an absolute reduction of speed in smoke within a certain range, Δ, of speeds around the average, i.e. speed reduction is independent from the initial walking speeds.

The five correlations generate different equations in relation to the type of data-set employed (current evacuation models either embed the data-sets by Jin [2008] or Frantzich and Nilsson [2004]) and the specific type of curve employed by the model developers (linear, non-linear, etc.).

The verification test proposed by Korhonen and Hostikka [2009] is modified to take into account of the different types of correlation employed by the models to represent the impact of reduced visibility (smoke) on occupant walking speeds.

Geometry
A corridor 2 m wide and 100 m long. One exit (1 m wide) is placed at the end of the corridor.

Scenario
Smoke reduces the walking speed due to the reduced visibility. The unimpeded walking speed of an occupant for a smoke-free environment is set to a constant value equal to 1.25 m/s. A constant extinction coefficient equal to 1.0 /m is implemented in the corridor prior to running the simulation. No external sources of lights are present in this test, i.e, the environment is assumed to be constituted only by objects which do not emit light. The occupant has to reach the exit at the end of the corridor.

Expected result
The expected result is that the time needed by the occupant to cover the distance of the corridor is the same as the time manually calculated employing the correlation used by the model (i.e. in line with the speed reduction factor used by the model).

Test method
The test method is the verification of model assignment. A quantitative evaluation of model results in terms of time differences is performed. In relation to the type of correlation employed by the model, the tester needs to identify a suitable quantitative method to evaluate the differences among the simulated and expected time.

<u>User actions</u>

The test should be repeated in order to verify different values in the correlation, i.e. different combinations of unimpeded walking speeds for a smoke-free environment and constant extinction coefficients needs to be tested. Examples of such values may be 1.0 m/s, 0.75 m/s, 0.5 m/s, and 0.25 m/s for the unimpeded walking speeds and 10/m, 7.5/m, 3.0/m, and 0.5/m for the extinction coefficient. These values are suggested in order to cover the range of walking speeds and extinction coefficients included in the two main data-sets available in the literature [Ronchi et al., 2013a]; i.e., Frantzich and Nilsson [2004] and Jin [2008]. It should be noted that the tester needs to know the correlation employed by the model and then compare the test results with hand calculations performed beforehand, i.e. the tester calculates in advance the assumed reduction of speed due to the smoke. Models may also consider the impact of smoke irritancy on people performance. This test does not consider the effects of irritant smoke and toxic gases on occupant speed (i.e. crawling behaviours, etc.), i.e., only the impact of reduced visibility on walking speed is taken into account. Kuligowski et al. [2010] highlighted that evacuation models may not include a sub-model simulating the impact of smoke on walking speeds. In that case, a reduced speed can be implicitly implemented within evacuation models, but it is not possible to simulate the impact of the change of visibility conditions over time on walking speeds. The tester should describe this limitation of the model under consideration.

Verif.2.6. Occupant incapacitation

A test is proposed to qualitatively and quantitatively verify the ability of evacuation models to simulate occupant incapacitation due to the toxic and physical effects of smoke. The incapacitation of building occupants is implemented in all evacuation models which attempt to represent the presence of smoke [Kuligowski et al., 2010] using the Fractional Effective Dose (FED) concept [Purser, 2008]. The suggested test is a modified version of the test designed by Korhonen and Hostikka [2009].

<u>Geometry</u>

A room with no fire source (10 m x 10 m x 3m).

<u>Scenario(s)</u>

The implementation of the FED concept is tested. Step 1: place an occupant in the centre of the room (see Figure 2). The occupant is held in a fixed initial position by setting a high pre-evacuation time (>10000000 s). Hazardous conditions are implemented in the model in relation to the incapacitation sub-model in use. Examples of such conditions are the exposure to toxic, irritant and physical hazards such as HCN, CO, CO_2, HCl, HBr, HF, SO_2, NO_2, elevated temperature, thermal radiation, etc. Step 2: Construct the same room and perform a FED measurement in the same location of the occupant, (either using hand calculations or an independent validated fire model using the same FED calculations implemented in the evacuation model).

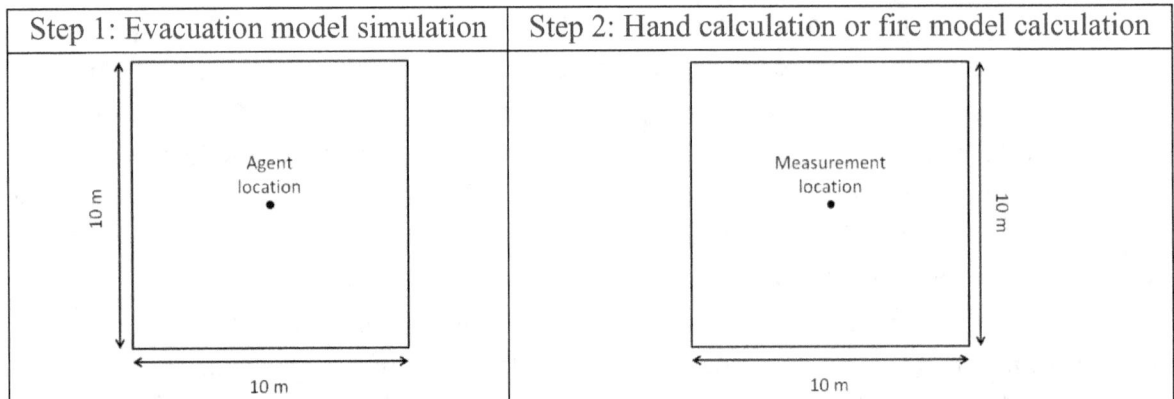

| Step 1: Evacuation model simulation | Step 2: Hand calculation or fire model calculation |

Figure 2. Geometric layout of Verif.2.6 test. On the left there is the geometry of the simulation test and on the right, the same geometry is employed for the measurements based on hand calculations or a fire model simulation.

Expected result

The expected result is that the time to reach occupant incapacitation (FED=1) in Step 1 is the same as the time to reach FED=1 in the measurement point in Step 2. This test should be repeated for each hazardous condition available in the incapacitation sub-model (e.g. CO or HCN concentrations, elevated temperature, etc.)

Test method

The test method employed is a quantitative verification of model assignment. The evaluation of the differences in the times to reach FED=1 during the two steps of the test is performed.

User actions

It should be noted that the tester needs to know the toxicity and hazard sub-model(s) embedded in the evacuation model to perform the test. The present test is a static test. Model testers may consider expanding the verification of FED calculations by considering an occupant moving in the space. If the model under consideration does not embed toxicity and hazard sub-models, it is recommended that the tester discusses this limitation in the documentation associated with the V&V of the model.

Verif.2.7. Elevator Usage

Current building codes are gradually implementing the use of elevators as a possible egress component [International Code Council, 2012] and an integral part of different egress strategies [Ronchi et al., 2013b]. Therefore, a test is suggested to verify the capability of evacuation models in simulating evacuation using elevators. It should be noted that elevators are gradually being implemented in evacuation models (10 out of 26 models in the review by Kuligowski et al. [2010].

Geometry

Construct two rooms, namely room 1 and room 2, placed at different heights having a floor-to-floor inter-distance equal to 3.5 m (see Figure 3 and Figure 4). Place an elevator connecting the two rooms in accordance to Figure 3 and Figure 4. Insert a 1 m wide exit in room 1.

Scenario

Insert an occupant having an unimpeded walking speed of 1 m/s in room 2 (See Figure 3 and Figure 4) with an instant response time. The elevator is the only egress component available for evacuation. The elevator has to start from room 1, reach room 2 and pick the occupant and then go back to room 1 to discharge the occupant. The tester has to define the kinematic settings for the elevator (e.g., elevator speeds, acceleration, open and close time, etc.).

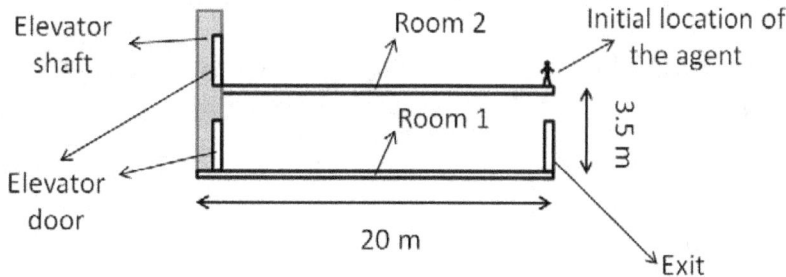

Figure 3. Schematic geometric layout of Verif.2.7. Test. Side view of the layout.

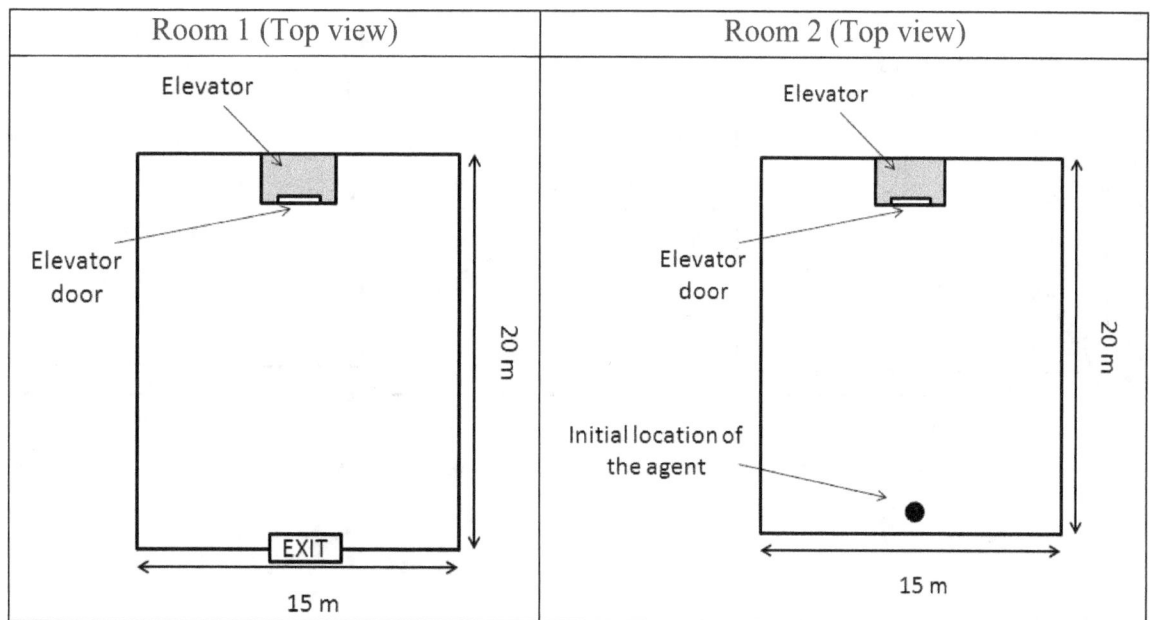

Figure 4. Schematic geometric layout of Verif.2.7. Test. Top view of room 1 and room 2.

Expected result

The expected result is that the occupant first enters the elevator in room 2. The same occupant is then discharged in room 1 and reaches the exit in room 1. If possible, this evaluation can be performed using the visualization tool of the model.

Test method

The test method is a qualitative verification of model assignment, i.e. the ability of the model to simulate evacuation using elevators.

User actions

If the model under consideration does not include an elevator sub-model, the tester is recommended to discuss this limitation in the documentation associated with the V&V of the model.

Verif.2.8. Horizontal counter-flows (rooms)

A test is suggested for the verification of the ability of models to simulate counter-flow. This test is a modified version of the IMO Test 8 and it is a verification of emergent behaviours concerning counter-flow.

Geometry

Two rooms 10 m (wide and long) connected via a corridor 10 m long and 2 m wide starting and ending at the centre of one side of each room [see Figure 5].

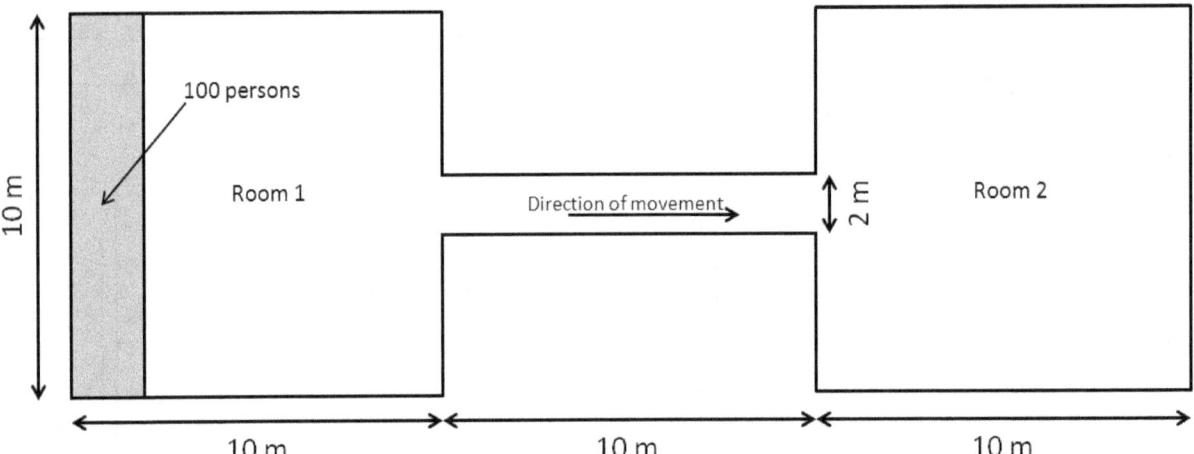

Figure 5. Geometric layout of Verif.2.8 Test based on the IMO Test 8 [International Maritime Organization, 2007].

Scenario

Choose a sub-population consisting of a population of 100 persons with response time equal to 0 s and distribute the walking speeds in accordance with the population of the building(s) (see

26

Lord et al. [2005] for possible occupant demographics). Step 1: One hundred persons move from room 1 to room 2, where the initial distribution is such that the space of room 1 is filled from the left with maximum possible density. The time the last person enters room 2 is recorded. Step 2: Step one is repeated with an additional ten, fifty, and one hundred persons in room 2. These persons should have identical characteristics to those in room 1. Both sub-populations move simultaneously to the opposite room and the time for the last persons from room 1 to enter room 2 is recorded.

Expected result
The expected result is that the recorded time increases as the number of persons in counter-flow increases.

Test method
The test method is a qualitative evaluation of the capabilities of the model of reproducing horizontal counter-flows (counter-flows in rooms). Model results need to be compared and the differences (expressed in terms of evacuation times) between the steps of the test are presented.

User actions
The model tester should qualitatively discuss the extent of the recorded time increases due to counter-flows.

Verif.2.9. Group Behaviours

Evacuation models [Kuligowski et al., 2010] often include the possibility to simulate the interactions between building occupants, i.e. group behaviours. In this instance, group behaviours only refer to occupant movement (i.e. they do not include decision-making, communication, etc.). This test is designed to perform a qualitative verification of the emergent behaviours of groups. This test identifies whether a group sub-model is available and if it is able to reproduce group behaviours not only as a set of individuals with the same characteristics, but as a group of occupants remaining together even in the case of different occupant characteristics (e.g., different occupant walking speeds).

Geometry
A room of size 15 m by 20 m with a 1 m exit.

Scenario
Five occupants are assigned to the same group of able-bodied adults, namely Group 1, in the top of the room (see zone 1 in Figure 6) with response time equal to 0 s. Four of the occupants of Group 1 have a constant unimpeded walking speed of 1.25 m/s. The fifth occupant of Group 1 has a constant unimpeded walking speed of 0.5 m/s. In the central part of the room 10 slower

able-bodied adult occupants, namely Group 2, with a constant unimpeded walking speed of 0.2 m/s are uniformly distributed in Zone 2 as it is shown in Figure 6. The occupants in Zone 1 have to reach the exit of the room.

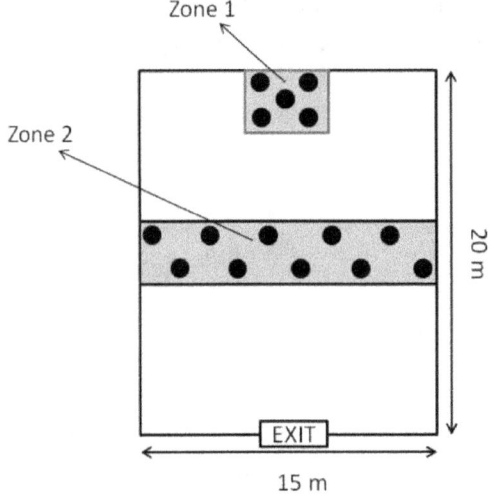

Figure 6. Schematic geometric layout of Verif.2.9. Test.

Expected result

The test should demonstrate that the occupants of Group 1 will reach the exit together (i.e., the times for occupants of Group 1 to reach the exit should not differ of more than 10 s). If possible, this evaluation can be performed using the visualization tool of the model. The choice of 10 s is arbitrary driven by the need to set a number to make a quantitative comparison. Preliminary tests were performed with an evacuation model which uses assumptions very similar to most of the models representing group behaviours in order to assess the approximate time needed to reach the exit and evaluate the expected differences.

Test method

The test method is an evaluation of emergent behaviours which uses quantitative criteria. The analysis is performed by comparing the time needed by the occupants of Group 1 to reach the exit.

User actions

If the model under consideration does not permit the simulation of group behaviours, the tester is recommended to discuss this limitation in the documentation associated with the V&V of the model.

Verif.2.10 People with movement disabilities.

People with disabilities are an important part of the population that model developers are starting to include in evacuation models. This test is designed for the verification of emerging behaviours of people with disabilities. Verif 2.10. is aimed at testing the possibility of simulating an occupant with reduced mobility (e.g. decreased travel speeds and increased space occupied by the occupants) as well as representing the interactions between impaired individuals and the rest of the population and the environment.

Geometry
Construct two rooms at different heights, namely room 1 (1 m above the ground level) and room 2 (at ground level), connected by a ramp (or a corridor/stair if the model does not represent ramps). Insert one exit (1 m wide) at the end of room 2 (see Figure 7 for the schematic representation of the rooms).

Scenarios
Scenario 1: Room 1 is populated with a sub-population consisting of 24 occupants in zone 1 (with an unimpeded walking speed of 1.25 m/s and the default body size assumed by the model) and 1 disabled occupant in zone 2 (the occupant is assumed to have an unimpeded walking speed equal to 0.8 m/s on horizontal surfaces and 0.4 on the ramp (see Figure 7). The disabled occupant is also assumed to occupy an area bigger than half the width of the ramp (>0.75 m) (e.g., a wheelchair user)[3]. All occupants have to reach the exit in room 2.

Scenario 2: Re-run the test and populate zone 2 with an occupant having the same characteristics of the other 24 occupants in zone 1 (i.e. no disabled occupants are simulated). All occupants have to reach the exit in room 2.

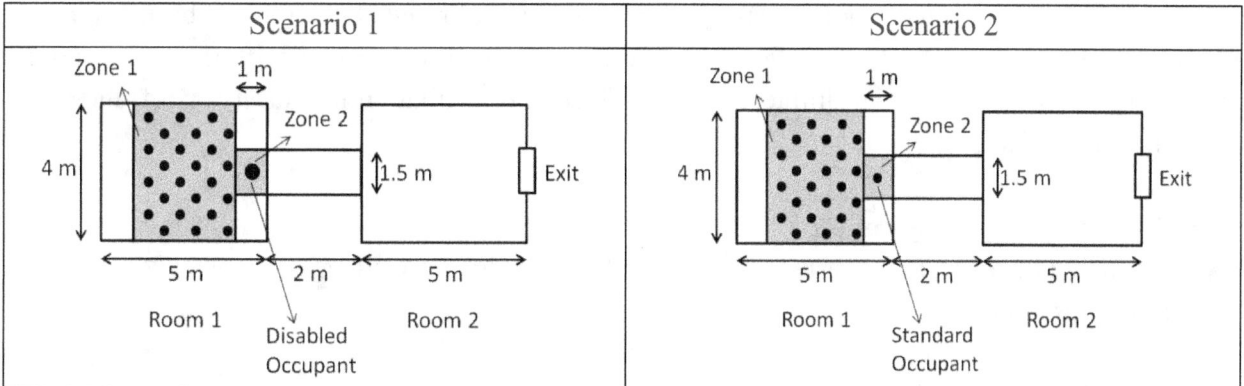

Figure 7. Schematic geometric layout of Verif.2.10. Test.

---

[3] It should be noted that this test is designed for models which permit the simulation of agents of different dimensions (i.e. continuous models or cellular automata models which allows the simulation of agents occupying more than one cell).

The expected result is that occupants in zone 1 in Scenario 1 reach the exit in a time slower than occupants in zone 1 in Scenario 2. If possible, this evaluation can be performed using the visualization tool of the model.

Test method

The test is a qualitative verification of emergent behaviours. The tester should qualitatively evaluate if the model is able to simulate disabled populations and their possible impact on the evacuation times.

User actions

If the model under consideration does not permit the simulation of people with movement disabilities or it does not permit the simulation of agents of different dimensions, the tester is recommended to discuss this limitation in the documentation associated with the V&V of the model.

3.1.3 Exit choice/usage

Tests may be provided to study either the ability of the user to specify exit use or the ability of the model to allocate exit use given certain parameters. Exit choice sub-models available in building evacuation models may rely on simple criteria (shortest distance, user-defined), allowing for a deterministic rather than predictive result. For the case of models based on deterministic criteria, it is expected that the occupants will always choose the closest exit in all scenarios if the exit choice is not driven by user input. An exit route allocation test based on IMO Test 10 is suggested. Two verification tests aimed at evaluating the capabilities of evacuation models in simulating social influence (Verif.3.2) and affiliation/familiarity with the exit (Verif.3.3) are also presented.

Verif.3.1. Exit Route Allocation
An exit route allocation is suggested in order to verify the deterministic assignment of exit usage. The test is based on IMO Test 10.

Geometry

Construct a corridor section with rooms in accordance with Figure 8.

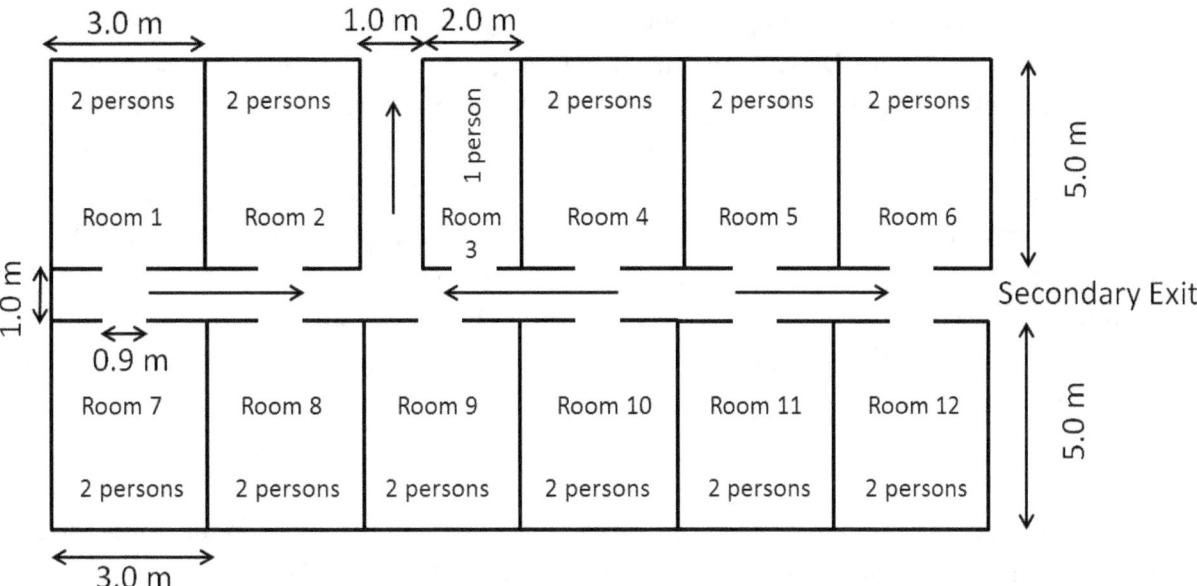

Figure 8. Schematic geometric layout of Verif.3.1. Test based on IMO test 10 [International Maritime Organization, 2007].

Scenario
Populate the rooms with occupants having walking speeds and characteristics in accordance with the expected demographics of the population of the building(s) (see Lord et al. [2005] for possible occupant demographics). Distribute the walking speeds and response times equal to 0 s over a population of 23 persons. The persons in room 1, 2, 3, 4, 7, 8, 9, and 10 are allocated to the main exit. All the remaining passengers are allocated the secondary exit.

Expected result
The allocated occupants move to the appropriate exits. If possible, this evaluation can be performed using the visualization tool of the model.

Test method
The test method is a qualitative verification of model assignment, i.e. the ability of the model to represent exit route allocation.

User actions
The tester needs to mention if the exit choice sub-model is based on deterministic assumptions or it is predictive in the documentation associated with the test where the results of the model are presented.

Verif.3.2. Social influence

One of the main factors that may impact route usage/exit choice is social influence [Deutsch and Gerard [1955], Latane' and Darley [1970], Nilsson and Johansson, 2009]. Social influence is defined as changes in attitudes, beliefs, opinions or behaviour as a result of the fact that one has encountered others [Hewstone and Martin, 2008]. An ideal test is suggested for the analysis of emergent behaviours regarding social influence in building evacuation models. This test is aimed at qualitatively verifying model capabilities to simulate the impact of social influence on exit choice. Previous studies demonstrated the importance of social influence as a key aspect that needs to be addressed to perform exit usage predictions [Kinateder, 2013]. This test requires an exit choice sub-model which includes the possibility of simulating social interactions and their impact on exit usage.

Geometry
Construct a room of size 10 m by 15 m. Two exits (1 m wide) are available on the 15 m walls of the room and they are equally distant from the 10 m long wall at the end of the room (see Figure 9, where the centre of the doors is 12 m from the 10 m long wall).

Scenarios
Scenario 1: Insert one occupant (occupant 1) in the room with a response time equal to 0 s and a constant walking speed equal to 1 m/s as shown in Figure 9 (the black dot represents the occupant which is 1 m away from the bottom wall that is 10 m long). The occupant does not have a preferred exit (i.e. they are not familiar with any of the exit). The occupant should be placed always in the same position among different runs and his/her position should be equidistant to both exits. Run the test several times until you get a stable percentage of exit usage for both exits i.e., exit usage does not vary more than 1 %[4] with an additional run. Annotate the exit usage for the two exits.

Scenario 2: Insert an additional occupant (occupant 2) in the room with an instant response time and a constant walking speed equal to 1 m/s as shown in Figure 9 (two occupants in total). The additional occupant is placed 2 m away from the bottom wall that is 10 m long. This occupant is deterministically assigned to Exit 2. Run the test several times until you get a stable percentage of exit usage for the two exits for both occupants i.e., the exit usage does not vary more than 1% with an additional run. Annotate the exit usage for both occupants.

---

[4] Different methods can be adopted to evaluate the convergence of the percentage of exit usage. The 1% requirement has been selected since it is deemed to be easily applied by model testers and it allows a comparison of percentages removing the confounding factor of the impact of the number of runs over exit usage.

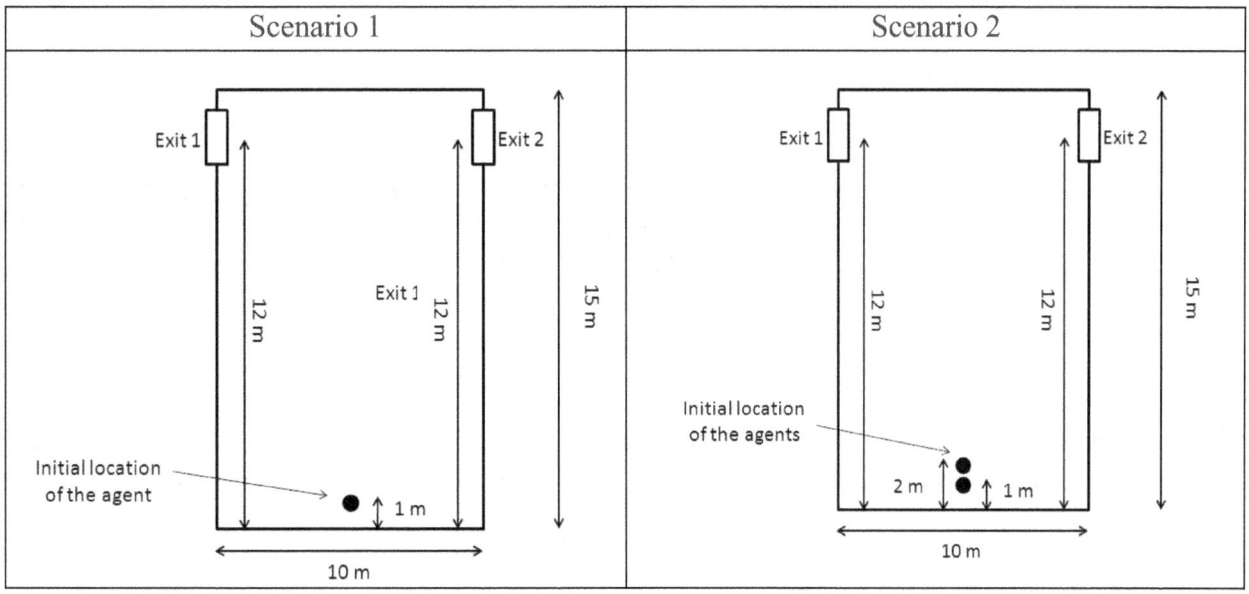

Figure 9. Schematic top view of the geometric layout for Verif.3.2. test.

Expected result

The expected result is that the usage of exit 2 is increased in scenario 2 for occupant 1.

Test method

The evaluation method of this test is a quantitative evaluation of model results in terms of exit usage.

User actions

It should be noted that the exit choice sub-models of evacuation models may rely on simpler criteria (shortest distance, user defined), i.e. they may be based on a deterministic choice of the user rather than a prediction of the exit usage. For this type of model it is expected that the occupants will always choose the closest exit in all scenarios if the exit choice is not driven by the user input. The tester needs to document this limitation.

Verif.3.3. Affiliation

This test is aimed at qualitatively verifying the capabilities of evacuation models to simulate the effect of an individual's familiarity with an exit on exit usage. This test belongs to the category of the verification tests of emergent behaviours (EB_VERIF). Affiliation is a concept introduced by Sime [1984], which relates to the likelihood of a person preferring the use of a familiar exit over an unfamiliar one (e.g. preferring to go towards the location employed to enter the building) during the evacuation process. This test requires an exit choice sub-model which includes a variable that can directly simulate the affiliation of the occupants with the exits. Several evacuation models may include algorithms which explicitly represent the influence of affiliation upon the decision-making process.

Also in this case, the tester should mention if the exit choice sub-algorithm of the evacuation model under consideration is based on deterministic criteria, i.e., if exit choice is only driven by distance criteria or user input. In those cases, this test is not considered as a verification of emergent behaviours but it represents analytical verification (i.e. a verification of model assignment).

Geometry
Construct a room of size 10 m by 15 m. Two exits (1 m wide) are available on the 15 m walls of the room and they are equally distant from the 10 m long wall at the end of the room (see Figure 10).

Scenarios
Scenario 1: Insert an occupant in the room with a response time equal to 0 s and a constant walking speed equal to 1 m/s as shown in Figure 10 (the black dot represents the occupant which is 1 m away from the 10 m long wall on the bottom of Figure 10). The occupant should always be placed in the same position among different runs and his/her position should be equidistant to both exits. The occupant is assumed to be unfamiliar with the exits. Run the test several times until you get a stable percentage of exit usage for both exits i.e., exit usage does not vary more than 1% with an additional run. Annotate the exit usage for the two exits

Scenario 2: Insert an occupant in the central area at the beginning of the corridor with an instant response time and a constant walking speed equal to 1 m/s as shown in Figure 10. This occupant is affiliated with Exit 2. The same occupant is not affiliated with Exit 1 (e.g. Exit 2 is the favoured exit chosen by the occupant if all the other conditions affecting choice are the same for all exits). Run the test several times until you get a stable percentage of exit usage for both exits i.e., exit usage does not vary more than 1% with an additional run[5]. Annotate the exit usage for both exits.

---

[5] Different methods can be adopted to evaluate the convergence of the percentage of exit usage. The 1% requirement has been selected since it is deemed to be easily applied by model testers and it allows a comparison of percentages removing the confounding factor of the impact of the number of runs over exit usage.

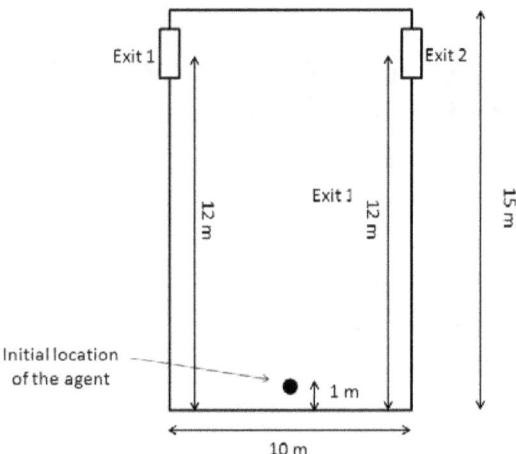

Figure 10. *Schematic top view of the geometric layout for Verif.3.3. test.*

<u>Expected result</u>

The expected result is that the usage of exit 2 in scenario 2 is higher than the exit 2 usage in scenario 1.

<u>Test method</u>

The evaluation method of this test is a quantitative evaluation of model results in terms of exit usage.

<u>User actions</u>

The model tester should document if the model includes a dedicated algorithm for the simulation of affiliation and if the exit choice sub-model is based on deterministic assumptions (i.e. user defined percentage of exit usage) or if it includes a predictive sub-algorithm.

3.1.4   Route Availability

This core element deals with the routes available to evacuees [Gwynne et al., 2012a]. A verification test (Verif.4.1.) is suggested in order to check the ability of the model to assign certain routes/egress components to occupants and modify route status over time (dynamic availability). For instance, a door can be rendered unavailable (over time) because of smoke, heat, etc. The test should verify that the model assignment is correct.

Verif.4.1. Dynamic availability of exits

This test is aimed at qualitatively evaluating the capabilities of the model to represent the dynamic availability of exits.

<u>Geometry</u>

Construct a room of size 10 m by 15 m. Two exits (1 m wide) are available on the 15 m walls of the room and they are equally distant from the 10 m long wall at the end of the room (see Figure 11).

<u>Scenario</u>

Insert an occupant in the room with a response time equal to 0 and a constant walking speed equal to 1 m/s as shown in Figure 11. Exit 1 becomes unavailable after 1 s of simulation time. Check the exit usage for both Exit 1 and Exit 2.

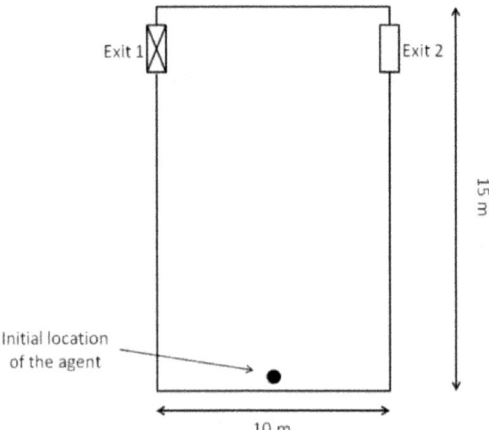

Figure 11. Schematic top view of the geometric layout for Verif.4.1. test.

<u>Expected result</u>

The expected result is that Exit 1 is not used by the occupant.

<u>Test method</u>

The model capabilities are analysed in this test using a quantitative evaluation of the results in terms of exit usage. If possible, this evaluation can be performed using the visualization tool of the model.

<u>User actions</u>

If the model does not include the possibility to simulate dynamic exit usage, the model tester should document this limitation.

3.1.5 Flow Constraints

This core behavioural element deals with the representation of the relationship between occupant speeds, flows, densities, the population size and the egress component under consideration [Gwynne et al., 2012a]. A verification test (Verif.5.1) is suggested to verify the capabilities to reproduce congestion within evacuation models. A test on maximum flow rates is also presented (Verif.5.2).

Verif.5.1. Congestion

A test is suggested for use to verify how the model simulates congestion. A modified version of the IMO Test 11 is proposed. The test is aimed at verifying the flow constraints in a staircase.

Geometry
Construct a room connected to a stair via a corridor (see Figure 12 for room, stair, and corridor dimensions).

Scenario
Populate the room with a sub-population consisting of 100 occupants, corresponding to a density[6] of 2.5 people/m^2, having the characteristics in accordance to the population of the building(s) (see Lord et al. [2005] for possible occupant demographics). Occupants have instant response times and walking speeds are distributed over a population of 100 persons.

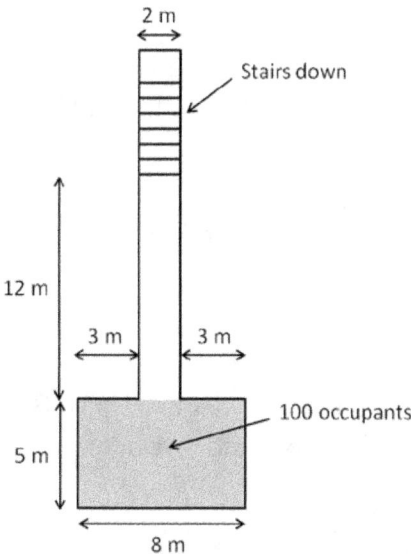

Figure 12. Schematic top view of the geometric layout for Verif.5.1. test IMO test 11 [International Maritime Organization, 2007].

---

[6] This high density has been chosen in order to investigate the case of congested areas, i.e. a relatively high number of occupants are placed in a narrow space.

Expected result

The expected result is that congestion appears at the exit from the room, which produces a steady flow in the corridor with the formation of congestion at the base (i.e. the bottom) of the stairs given the different flow characteristics of the corridor and the stair. If possible, this evaluation can be performed using the visualization tool of the model.

Test method

The test method is a qualitative verification of model results in terms of simulated congestions.

User actions

It should be noted that since building evacuations generally occur moving downward, the geometry of the IMO Test 11 has been modified, i.e., the stairs lead to a lower level rather than an upper level.

Verif.5.2. Maximum flow rates

A test is suggested to set a conservative requirement of maximum admitted flow rates. This test is based on the IMO Test 4.

Geometry

Construct a room of size 8 m by 5 m with a 1 m exit located centrally on the 5 m wall.

Scenario

Place 100 occupants in the room and assign them to the exit[7].

Expected result

The flow rate at the exit over the entire period should not exceed a pre-defined maximum threshold.

Test method

The test method is a quantitative evaluation of model results, i.e. the comparison between the results produced by the model and the maximum flow rate.

User actions

This test may also be susceptible to the type of grid/network in use in the case of fine and coarse network models. For this reason, the tester should demonstrate the sensitivity of the results in

---

[7] It corresponds to a density of 2.5 people/m². This high density has been chosen in order to investigate flow rates in the case of congested areas, i.e. a relatively high number of occupants are placed in a narrow space

relation to a different discretization of the space. This test can be interpreted in two different ways. First, it is a verification tests if the model under consideration represents flows through doors using restricted flows. However, it can be instead intended as an external validation requirement if flows are emergent and the tester wants to ensure that a maximum flow rate is not exceeded. An example of the maximum flow rate is the value recommended by the MSC/Circ. 1238 (1.33 p/m/s) [International Maritime Organization, 2007]. The model tester should document the assumptions adopted in the representation of the flows.

3.2   Validation Tests

This section presents a list of tests suggested for the validation of evacuation models. A set of considerations, listed here, is necessary before the discussion on the tests suggested for validation:

1) Experimental data-sets on human behaviour in fire are scarce, thus limiting the possible number of validation tests that can be performed.
2) The definition of benchmark validation tests relies on the techniques adopted to collect evacuation experimental data (and the subsequent uncertainties), the documentation provided with the experimental data-sets, and their availability to the public.
3) Current evacuation models, given the lack of human behaviour data, are relatively limited in terms of behavioural predictions, i.e., they are mostly deterministic or user dependent.
4) The validation tests are chosen in order to increase the understanding of evacuation model limitations.

A comprehensive list of validation tests needs to include experimental data-sets relating to a full range of possible behaviours and scenarios representing the evacuation process. Nevertheless, the lack of experimental data-sets makes it difficult to validate all aspects of evacuation modelling tools.

Evacuation models may be created starting from a set of specific experimental data. The validation of a model should not be performed using only the same data used for its development. This would produce a circular logic that may limit the extent to which model predictions may be generalised for all possible scenarios.

An alternative type of evacuation model design relies on the use of hypothetical assumptions, rather than experimental data-sets. An example of this issue is the simulation of merging flows in staircase. For example, the evacuation community is currently debating the appropriate stair merging ratio to be adopted in tall buildings. Few experimental studies [Boyce et al., 2012], Hokugo et al. [1985], Melly et al. [2009], Fang et al., 2012] are available on the topic. For this reason, it is not possible to provide validation tests for this issue or any other evacuation issues for which there is a lack of understanding of actual occupant behaviours.

To date, the definition of a complete set of experimental data to be used for the validation of the core behavioural components of evacuation models is not possible due to the limited amount of experiments suitable for validation. Nevertheless, this section suggests a set of examples of experimental/actual data-sets that are suitable for the validation of specific aspects of evacuation. Consequently, the examples provided should not be considered as an exhaustive list of evacuation model validation tests. As soon as further data-sets and theory are developed, the list of validation tests can be expanded and updated.

3.2.1. Examples of data-sets for model validation

This section presents a set of potential data-sets that may be suitable for the analysis of the core behavioural components of building evacuation models. As previously mentioned, this list should not be considered an exhaustive list of validation tests but rather as a set of examples of possible data-sets. Table 2 presents datasets suggested for the analysis of the core behavioural components of evacuation models. The selection of potential data-sets is based on their availability to the public, the documentation associated with them and the data collection/analysis method employed. Suggestions on the variables that can be used for the comparison between model predictions and the experimental data are also provided. Detailed information on the data-sets can be found in the corresponding references. Examples of the application of evacuation data-sets for the validation of evacuation models can be found in Ronchi et al. [2013b].

Table 2. Examples of possible experimental data for the validation of the main core components of building evacuation models.

Core component	Sub-element	Suggested variable for the comparison	Experimental data[8]
1	Pre-evacuation time distribution	Evacuation times, exit choice	Bayer and Rejno [1999]
2	Stairwell evacuation	Evacuation times	Kuligowski and Peacock [2010]
3,4	Impact of way-finding installations	Exit choice	Nilsson [2009]
5	Small scale experiment	Evacuation times, movement speeds, flows	Frantzich et al. [2007]
5	Small scale experiment	Evacuation times, movement speeds, flows	Hogeendorn and Daamen [2005]
5	Small scale experiment	Evacuation times, movement speeds, flows	Seyfried et al. [2007]
1,2,3,4,5	Full building evacuation	Evacuation times	The Station nightclub fire [Grosshandler et al., 2005]

---

[8] This data should not already be used by the model as benchmark for its development.

4.0 Uncertainty in Evacuation Modelling

The previous sections of this document provided a list of verification tests and suggested data-sets for validation of evacuation models. In order to provide an assessment of the capabilities of evacuation models, there is a need to discuss the uncertainty associated with evacuation modelling which is reflected in the methods to perform verification and validation.

Uncertainty is divided into different components in the context of fire safety engineering and modelling [Hamins and McGrattan, 2007]: *model input uncertainty, measurement uncertainty, and intrinsic uncertainty.*

1) Model input uncertainty is associated with the parameters obtained from experimental measurements that are used as model input, i.e. the assumptions employed to derive model input from the experiments as part of the model configuration process.
2) Measurement uncertainty is associated with the experimental measurement itself, i.e., the data collection techniques employed.
3) Intrinsic uncertainty is the uncertainty associated with the physical and mathematical assumptions and methods that are intrinsic to the model formulation.

To explain the different types of uncertainties in the case of building fire evacuation, we refer to the simulation of evacuation movement of a group of building occupants during a fire drill. The uncertainty associated with the measurement of the walking speeds is the measurement uncertainty. The approximated distribution used to configure the model input is the model input uncertainty. The uncertainty associated with the calculation method employed by the model to represent people movement is the intrinsic uncertainty.

In the case of evacuation modelling, uncertainty includes an additional component, here named *behavioural uncertainty*. Behavioural uncertainty is uncertainty associated with the stochastic factors employed to represent human behaviour [Averill, 2011], and a single experiment or model run may not be representative of a full range of the behaviours of the occupants. In fact, *"evacuate the same building with the same people starting in the same places on consecutive days and the answers could vary significantly"* [Averill, 2011]. There is a subsequent need for multiple experimental data-sets of the same scenario to understand the possible variability of occupant behaviours in each individual evacuation scenario. Unfortunately, experimental data-sets on human behaviour in fire are scarce and single data-sets are often the only available reference for the study of an individual scenario. Ideally, model testers should use a range of evacuations from the real world and when models produce outputs in line with the range of the real-world outcomes, the model is validated. Since data are scarce, model testers often rely on a single real-world observation without understanding, whether that single curve is representative of the average behavioural performance. Behavioural uncertainty needs to be analysed in both experimental and modelling studies. In this context, the assessment of the variability of simulation results in relation to *behavioural uncertainty* is a key issue to be discussed. This is reflected in the estimation of the convergence of an individual evacuation simulation scenario towards an "average" predicted occupant evacuation time-curve. The assessment of evacuation model results may also include the analysis of the tails of the distribution rather than the analysis of the peaks (i.e. average values). Nevertheless, the authors argue that the study of the average model predictions together with the variability of results around the average is deemed to be a useful method to analyse *behavioural uncertainty*. It should be noted that the term *behavioural uncertainty* is here introduced in the context of fire safety science, i.e. the term may have different meanings in other research fields.

Fire modellers and evacuation modellers treat uncertainty in different ways. Uncertainty is generally treated in fire models as a deterministic problem, i.e., it is traditionally studied by analysing the sensitivity of the model output in relation to the variability of the model input [Jones et al, 1995]. This is driven by the fact that fire models are generally based on deterministic equations (e.g., McGrattan et al. [2010], Jones et al. [2009]). On the other hand, evacuation models often treat uncertainty as a stochastic problem. The main difference between most of the fire and evacuation models is in the definition of the model input and algorithms. For instance, fire model input is generally based on a single input (e.g., a single curve for Heat Release Rate or single values for the characteristics of a burner, etc.). In evacuation models, inputs are generally inserted in terms of distributions of possible values. Also, the underlying algorithms of evacuation models often make use of probabilities. For instance, models may employ pseudo-casual algorithms to simulate the probability of an action to occur at each repeated run (for instance the exit that an agent will use may vary over different runs of the same scenario). The consequence is that, excluding the uncertainty associated with the modelling assumptions, in fire models, the uncertainty in the output is essentially driven by the choice of the values for the input (a single input will give a single output), while output uncertainty in evacuation models is due to the choice of the input as well as the underlying probabilistic algorithms inside the model (a single input may produce multiple results).

The use of stochastic/random variables is driven by the inability to confidently represent all cues and factors affecting human behaviour, which are reflected in different methods of agent representations (e.g., the use of random variables). There is a dual interpretation on why model developers adopt this solution. The first interpretation is that the "human element" introduces factors that are not entirely predictable. Another interpretation may instead rely on the fact that the current knowledge on human behaviour is limited and there may never be enough information to predict human response with any degree of certainty.

To address the uncertainty associated with human behaviour, evacuation models often employ distributions or stochastic variables to simulate people movement (Kuligowski et al. [2010], Lord et al. [2005], Ronchi and Kinsey [2011]), e.g., distribution of walking speeds, distribution of pre-evacuation times, etc. In fact, random numbers/seeds may be employed to solve space conflict resolution, simulate exit choice, familiarity with the exit, queuing behaviour, etc. When distributions are created adopting a random sampling method, This leads to the generation of multiple occupant-evacuation time curves for the same scenario using the same model inputs are produced. Random variables may be intrinsic of the model algorithms, and model users may not have control/access to them (especially in closed-source models). This leads to the need for a study of the variability of the results associated with the random variables embedded in the models.

Therefore, evacuation modellers face the problem of selecting the appropriate number of runs to be simulated in order to be representative of the average model outcome. This problem arises both during the use of evacuation models for a fire safety design as well as during validation studies. In fact, two main questions can be asked during the simulation of evacuation scenarios that include distributions or stochastic variables: 1) Which occupant-evacuation time curve is representative of model predictions in a fire safety design? 2) Which occupant-evacuation time curve should be used as reference during the comparison with experimental data in a validation study? To date, the answers to these questions are left as a qualitative judgment of the evacuation model user. For instance, in the context of evacuation model validation, model users may select the best model prediction during the comparison with experimental data [Galea et al., 2012b] or employ the model's average total evacuation time (possibly including information on the standard deviation) as representative of model predictions. The study of the average total evacuation times and their corresponding standard deviations provides insights only on the required safe escape time, rather than the whole evacuation process. There is instead a need for a method which investigates the size of the variation for the whole occupant-evacuation time curve. An alternative method is the simulation of a fixed number of repeated runs in order to consider the probabilistic nature of the evacuation process, as prescribed in the IMO Guidelines [International Maritime Organization, 2007]. Nevertheless, to date, there is no universally accepted quantitative method to estimate how the average prediction may vary over the number of runs.

In addition, complex evacuation scenarios may be computationally expensive to simulate. For instance, previous research on the use of distribution curves for Monte Carlo simulations for uncertainty analysis in evacuation model predictions have demonstrated the need for a large computational effort [Lord et al., 2005]. Therefore there is a need to optimize the selection of the number of runs of the same scenario in order to be representative of occupants' "average behaviour", and provide a quantitative and computationally inexpensive characterisation of the variability associated with the simulated runs (and a subsequent estimation of the behavioural uncertainty associated with an individual evacuation model setup).

The study of model predictions are generally performed using a statistical treatment of the data or a qualitative evaluation of the simulation results. Those methods both rely on user's expertise in terms of the selection of the statistical method to employ or the evaluation of the reliability model predictions. A method which permits a simple and computationally inexpensive analysis of model predictions is functional analysis. This branch of mathematics represents curves as vectors, and uses geometrical operations on the curves. Functional analysis operations are currently employed during the comparison of fire model evaluations and experimental data [Peacock et al., 1999], International Standards Organization, 2008] and the comparison between evacuation model results and experimental data [Galea et al., 2012b]. Nevertheless, functional analysis has not been employed so far to compare evacuation model predictions (produced by a single model or multiple models) against each other to analyse the uncertainty associated with the number of runs of the same evacuation scenario, i.e. behavioural uncertainty.

This section proposes a set of convergence criteria for the analysis of the variability of evacuation model predictions of the same evacuation scenario (i.e. the same model input which includes distributions or stochastic variables) in relation to the number of runs. A procedure for the definition of the optimal number of runs - in relation to the evacuation scenario, the model in use, and the scope of the simulations - is presented. The scope of the present work is to provide a quantitative method to assess the variability associated with the number of runs of the same evacuation scenario. The proposed method allows the analysis of behavioural uncertainty and the prediction of the average occupant-evacuation time curve in relation to pre-defined acceptance criteria.

4.1 A method for the study of behavioural uncertainty in evacuation modelling

The use of models adopting stochastic/random variables for the simulation of human behaviour creates the need for a systematic and quantitative analysis of behavioural uncertainty. To address this issue, this section presents a methodology for the analysis of behavioural uncertainty in evacuation modelling [Ronchi et al., 2013c]. It includes the definition of five convergence criteria for the analysis of the occupant-evacuation time curves produced by evacuation models and a procedure for the assessment of the optimal number of runs in relation to pre-defined acceptance criteria.

The proposed methodology is based on the definition of a set of convergence measures that sufficiently describe the distribution of occupant-evacuation time curves. This is addressed by

45

constructing a series for each measure and demonstrating that the measure is sufficiently close to the expected value, i.e. the series converge to the average occupant-evacuation time curve.

A series $S = \{s_i, ..., s_n\}$ converges to S_c if for any positive real value e there is an n such that $|S_c - s_n| < e$.

The series represents the evacuation time predictions of evacuation models based on sample data. This will imply that the series will likely not smoothly converge, meaning that it might happen that $|S_c - s_{n+1}| > |S_c - s_n|$. In order to increase the confidence that the series have sufficiently converged, a requirement that the last b values of the series (the convergence measures) are within S_c is added. For some series we might not know the expected value S_c, i.e., the value to which the series is convergent. In those cases, the last current value of the series is used as the best estimate of the value the series converges to.

4.1.1 Functional analysis concepts

Before discussion of convergence criteria, there is a need to introduce three concepts of functional analysis, namely the Euclidean Relative Difference (ERD), the Euclidean Projection Coefficient (EPC) and the Secant Cosine (SC). Initial applications of these concepts in the field of fire science are discussed by Peacock et al. [1999] and Galea et al. [2012b].

The single comparison of two individual points in a curve can be made by finding the norm of the difference between the two vectors representing the data. A norm represents the length of a vector. The distance between two vectors corresponds to the length of the vector resulting from the difference of the two vectors. For a generic vector $\vec{x}$, the norm is represented using the symbol $||\vec{x}||$. This concept can be extended to multiple dimensions. The distance between two generic multi-dimensional vectors $\vec{x}$ and $\vec{y}$ is therefore the norm of the difference of the vectors $||\vec{x} - \vec{y}||$. The Euclidean relative difference between two vectors can be normalized as a relative difference to the vector $\vec{y}$ (see Equation 6).

$$ERD = \frac{||\vec{x} - \vec{y}||}{||\vec{y}||} = \sqrt{\frac{\sum_{i=1}^{n}(x_i - y_i)^2}{\sum_{i=1}^{n}(y_i)^2}} \qquad \text{[Equation 6]}$$

The Euclidean Relative Difference (ERD) represents, therefore, the overall agreement between two curves.

Two components can be considered during the comparison of two vectors, namely the distance between two vectors and the angle between the vectors.

The concept of projection coefficient a is introduced. From a geometric point of view, the vector $a\vec{x}$ is the projection of the vector $\vec{y}$ onto the vector $\vec{x}$ (see Figure 13).

Figure 13. The projection coefficient for two vectors.

a defines a factor which reduces the distance between two vectors to its minimum (see Figure 13). The solution of the minimum problem is found and corresponds to Equation 7.

$$a = \frac{||\vec{y}||}{||\vec{x}||} \cos \beta \qquad \text{[Equation 7]}$$

$< x, y >$ is the inner product of two vectors, i.e., the product of the length of the two vectors and the cosine of the angle between them. The inner product can be interpreted as the standard dot product; producing Equation 8.

$$< \vec{x}, \vec{y} > = \sum_{i=1}^{n}(x_i y_i) \qquad \text{[Equation 8]}$$

The Euclidean Projection Coefficient (EPC) is found by studying the minimum problem, i.e., studying when the derivative of the function is zero (see Peacock et al. [1999] for the full solution of the minimum) and it corresponds to:

$$a = EPC = \frac{<\vec{x},\vec{y}>}{||\vec{y}||^2} = \frac{\sum_{i=1}^{n}(x_i y_i)}{\sum_{i=1}^{n} y_i^2} \qquad \text{[Equation 9]}$$

EPC defines a factor which when multiplied by each data point of the vector $\vec{y}$ reduces the distance between the vectors $\vec{y}$ and $\vec{x}$ to its minimum, i.e. the best possible fit of the two curves.

The concept of Secant Cosine (SC) is also introduced. It represents a measure of the differences of the shapes of two curves. This is investigated by analysing the first derivative of both curves. For n data points, a multi-dimensional set of n-1 vectors can be defined to approximate the derivative. This produces Equation 10 [Peacock et al., 1999]:

$$SC = \frac{<\vec{x},\vec{y}>}{||\vec{x}||||\vec{y}||} = \frac{\sum_{i=s+1}^{n}\frac{(x_i-x_{i-s})(y_i-y_{i-s})}{s^2(t_i-t_{i-1})}}{\sqrt{\sum_{i=s+1}^{n}\frac{(x_i-x_{i-s})^2}{s^2(t_i-t_{i-1})}\sum_{i=s+1}^{n}\frac{(y_i-y_{i-s})^2}{s^2(t_i-t_{i-1})}}} \qquad \text{[Equation 10]}$$

Where:

t is the measure of the spacing of the data, i.e., $t = 1$ if there is a data point for each occupant;

47

s represents the number of data points in the interval;

n is the number of data points in the data-set.

When the Secant Cosine is equal to unity, the shapes of the two curves are identical. Depending on the value for *s*, the noise of the data is smoothed out. An example of the impact of different values of *s* on the SC is shown in Figures 14 and 15. Figure 14 shows two hypothetical curves (obtained by 120 values for *x* and *y* corresponding to 120 arbitrary data-points) which include noise or no noise. The comparison between the shapes of the two curves is made using different *s* values. For instance, Figure 14 shows that if s=1, all noise is taken into account in the calculation of the Secant Cosine applying Equation 10, while if s=60, the noise is smoothed out and the curved line is considered as a straight line in the calculation of the Secant Cosine. Figure 15 shows that the use of higher values for *s* reduces the impact of the noise in the comparison, i.e., the Secant Cosine tends to 1 in relation to an increase in the values for s.

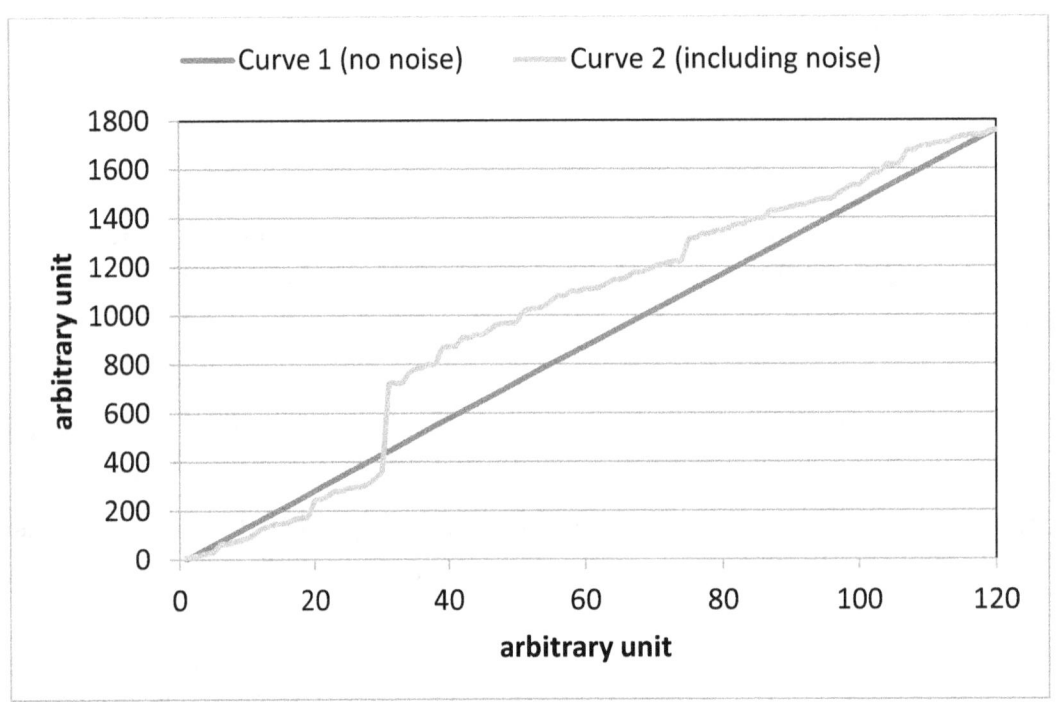

Figure 14. Hypothetical curves including noise (grey curve) and not including noise (black curve)

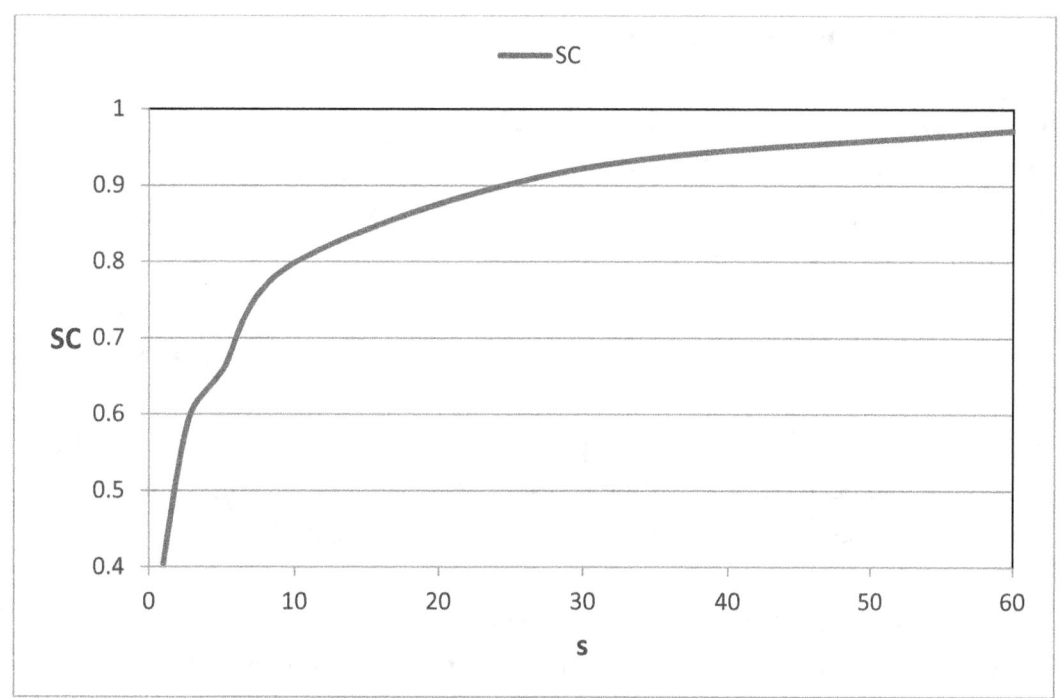

Figure 15. Secant Cosine in relation to different s values.

Nevertheless, *s* should not be too large, so that the natural variations in the data are kept. An example of this issue is provided in Figure 16, where, considering a hypothetical set of 4 data-points, different values for *s* generate either SC=1 for s=4 (the shape of the curves appear identical) or SC≠1 in the case of *s*=1 and *s*=2.

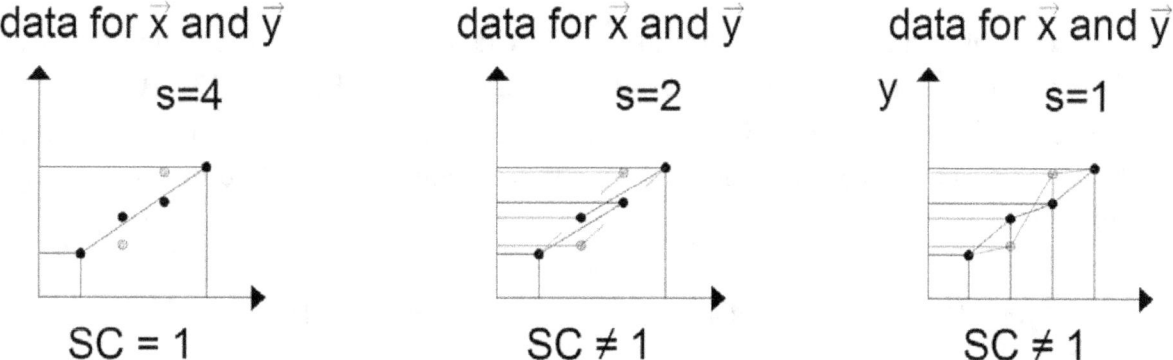

Figure 16. Schematic representation of the use of different values for s during the calculation of the Secant Cosine.

4.1.2 Convergence measures

A set of variables are introduced in order to present the method of analysis of evacuation model predictions based on functional analysis and convergence criteria. The measured experimental data are represented using vector $\vec{E}$ (see Equation 11), where E_i represents the measured evacuation time for the ith occupant.

$$\vec{E} = (E_1, \dots, E_n) \qquad \text{[Equation 11]}$$

For example, in the case of i=3 occupants, i.e., $\vec{E} = (E_1, E_2, E_3)$, E_1 is the measured evacuation time corresponding to occupant 1, E_2 is the measured evacuation time corresponding to occupant 2 and E_3 is the measured evacuation time corresponding to occupant 3.

The simulated times are represented by the vector $\vec{m}$ (see Equation 12), where m_i is the simulated evacuation time for the ith occupant, and m_n represents the evacuation time corresponding to the last occupant out of the building.

$$\vec{m} = (m_1, \dots, m_n) \qquad \text{[Equation 12]}$$

Therefore, $\vec{m} = (m_1, m_2, m_3)$, where m_1 is the simulated evacuation time corresponding to occupant 1, m_2 is the simulated evacuation time corresponding to occupant 2 and m_3 is the simulated evacuation time corresponding to occupant 3.

Several runs of the same scenarios are simulated. The simulated evacuation times of each occupant i in each jth run are represented using n vectors $\vec{m}_{ij}$ (see Equation 13). Here, q is the total number of occupants and n is the total number of runs. One assumption is that occupants are ranked in accordance to their evacuation time, i.e. occupants may evacuate the building in a different order in different runs.

$$\vec{m}_{ij} = (m_{11}, \dots, m_{ij}, \dots, m_{qn}) \qquad \text{[Equation 13]}$$

Considering nine runs of the same evacuation scenario including the same three occupants, 9 vectors $\vec{m}_{ij}$ are obtained where i=3 and j=9, i.e., $\vec{m}_{i1} = (m_{11}, m_{21}, m_{31}), \vec{m}_{i2} = (m_{12}, m_{22}, m_{32}), \dots, \vec{m}_{i9} = (m_{19}, m_{29}, m_{39})$.

The next variable that is presented is associated with the calculation of the arithmetic mean of the values of the runs. The jth average curve of evacuation times produced by the model considering the arithmetic mean of the values of all runs is represented using an n dimensional vector $\overrightarrow{M}_j$ (see Equation 14), where $M_1 = \frac{1}{n}\sum_{j=1}^{n} m_{1j}$, $M_2 = \frac{1}{n}\sum_{j=1}^{n} m_{2j}$, ..., $M_n = \frac{1}{n}\sum_{j=1}^{n} m_{qj}$

$$\overrightarrow{M}_j = (M_1, ..., M_j, ..., M_n) \qquad\qquad \text{[Equation 14]}$$

Considering the previous example, i.e. 3 occupants and 9 runs ($i=3$ and $j=9$), the average curve $\overrightarrow{M}_1$ corresponds to the values of the first run. The average curve for a sub-set of 4 runs will generate $\overrightarrow{M}_4$ which corresponds to the arithmetic means of the values up to the fourth run. In the case of all 9 runs, $\overrightarrow{M}_9$ corresponds to the arithmetic means of the values of all runs.

Figure 17 presents vector $\overrightarrow{M}_j$ in relation to the number of runs under consideration.

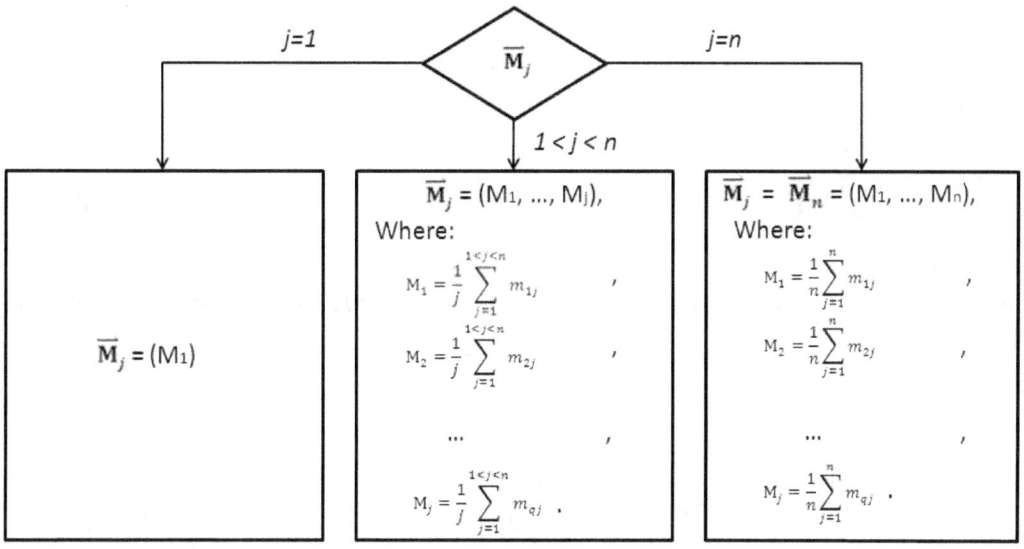

Figure 17. Vector $\overrightarrow{M}_j$ in relation to the number of runs.

Hence, if $j=1$, $\overrightarrow{M}_j=(M_1)$, i.e. the average curve corresponds to the curve of the first run. If $1 < j < n$, $\overrightarrow{M}_j$ becomes $\overrightarrow{M}_j=(M_1, ..., M_j)$ where $M_1 = \frac{1}{j}\sum_{j=1}^{1<j<n} m_{1j}$, $M_2 = \frac{1}{j}\sum_{j=1}^{1<j<n} m_{2j}$, ..., $M_j = \frac{1}{j}\sum_{j=1}^{1<j<n} m_{qj}$. $\overrightarrow{M}_j$ represents then the average curve corresponding to $1 < j < n$ runs. Considering 4 vectors $\overrightarrow{m}_{ij}$ corresponding to the predicted evacuation times for three occupants for $j=4$ of $n=9$ runs, $\overrightarrow{M}_4=(M_1 = \frac{1}{4}\sum_{j=1}^{1<4<9} m_{1j}, M_2 = \frac{1}{j}\sum_{j=1}^{1<4<9} m_{2j}, M_3 = \frac{1}{4}\sum_{j=1}^{1<4<9} m_{3j})$.

If $j=n$, $\overrightarrow{M_j}$ becomes $\overrightarrow{M_n}=(M_1, \ldots, M_n)$ where $M_1 = \frac{1}{n}\sum_{j=1}^{n} m_{1j}$, $M_2 = \frac{1}{n}\sum_{j=1}^{n} m_{2j}$, ..., $M_n = \frac{1}{n}\sum_{j=1}^{n} m_{qj}$. Thus, $\overrightarrow{M_n}$ represents the average curve corresponding to all $j=n$ runs. For instance, if $n=9$ runs, $\overrightarrow{M_9}=(M_1 = \frac{1}{9}\sum_{j=1}^{9} m_{1j}, M_2 = \frac{1}{9}\sum_{j=1}^{9} m_{2j}, M_3 = \frac{1}{9}\sum_{j=1}^{9} m_{3j})$.

Convergence measure 1: Total Evacuation Time (TET)

The vector m_n can also be called TET_j, total evacuation time (also called Required Safe Egress Time in the context of performance based design [Purser and Bensilum, 2001]), corresponding to the jth run. Therefore, there are several simulated TET_j, each one corresponding to the jth run for a total of n runs.

The jth total evacuation times TET_i for n runs of the same scenario simulated with an evacuation model can be represented using the vector $\overrightarrow{TET} = (TET_1, \ldots, TET_n)$.

The arithmetic mean of the total evacuation times for j runs can be expressed using TET_{avj} (see Equation 15):

$$TET_{avj} = \frac{1}{j}\sum_{i=1}^{j} TET_i \qquad \text{[Equation 15]}$$

The set of all n consecutive mean total evacuation times TET_{avj} of the same scenario simulated with an evacuation model is $TET_{av} = (TET_{av1}, \ldots, TET_{avn})$. TET_{av1} is assumed to correspond to the value in run 1, TET_{av2} is the average for $j=2$, ..., TET_{avn} is the average for $j=n$.

Applying the law of large numbers, the consecutive mean total evacuation times TET_{avi} can be interpreted as a series converging to an expected value (the mean total evacuation time). Hence, a measure of the convergence of the series can be performed.

A measure of the convergence of two consecutive mean total evacuation times TET_{avj} (e.g. TET_{av1} and TET_{av2}) is obtained calculating TET_{convj} (see Equation 16). It is expressed (in %) as the difference of two consecutive mean total evacuation times divided by the last mean evacuation time. This convergence measure assumes that the best approximation of the expected value (the mean total evacuation time) is the last mean evacuation time. This measure is useful to evaluate the impact of an additional run on the average predicted total evacuation time. This produces a total of $p=n-1$ TET_{convj}.

$$TET_{convj} = \left| \frac{TET_{avj} - TET_{avj-1}}{TET_{avj}} \right| \qquad \text{[Equation 16]}$$

The last TET_{convj} value, corresponding to all n runs is $TET_{convFIN}$ (see Equation 17).

52

$$TET_{convFIN} = \left| \frac{TET_{avp} - TET_{avp-1}}{TET_{avp}} \right|$$
[Equation 17]

Convergence measure 2: Standard Deviation (SD) of total evacuation times

Assuming a normal distribution of total evacuation times, convergence variables can also be presented in terms of the standard deviation of total evacuation times.

The jth standard deviation SD_j for n runs of the total evacuation time of the same scenario simulated with an evacuation model can be represented by the vector $\overrightarrow{SD} = (SD_1, ..., SD_n)$.

Also in this case, the application of the law of large numbers permits the interpretation of the consecutive standard deviations of total evacuation times SD_j as a series convergent to an expected value (the mean standard deviations of total evacuation times). Therefore, a measure of the convergence of the series is possible.

A measure of the convergence of two consecutive standard deviations SD_j (e.g. SD_1 and SD_2) is obtained by calculating SD_{convj}. It is expressed (in %) as the difference of two consecutive standard deviations divided by the last standard deviation (see Equation 18). This produces a total of $p=n-1$ SD_{convj}. This convergence measure assumes that the best approximation of the expected value (the mean standard deviation of total evacuation times) is the last standard deviation of total evacuation times. This measure is useful to evaluate the impact of an additional run on the standard deviation.

$$SD_{convj} = \left| \frac{SD_j - SD_{avj-1}}{SD_j} \right|$$
[Equation 18]

The last SD_{convj} value, corresponding to all n runs, is $SD_{convFIN}$ (see Equation 19).

$$SD_{convFIN} = \left| \frac{SD_{avp} - SD_{avp-1}}{SD_{avp}} \right|$$
[Equation 19]

Convergence measure 3: Euclidean Relative Difference (ERD)

A set of Euclidean Relative Differences (ERD) can be calculated, each one corresponding to two consecutive pairs of vectors $\overrightarrow{M_j}$ representing the progressive average occupant-evacuation time curves.

A vector $\overrightarrow{ERD} = (ERD_1, ..., ERD_p)$ is made of p consecutive ERD_j where $p=j-1$, corresponding to an average j runs of the same scenario simulated with an evacuation model. For instance, in the case of $j=4$ runs, $\overrightarrow{ERD} = (ERD_1, ERD_2, ERD_3)$ where ERD_1 is calculated from the comparison between M_1 and M_2, ERD_2 is calculated from the comparison between M_2 and M_3 and ERD_3 is calculated from the comparison between M_3 and M_4. M_1 represents the curve from run 1, M_2 represents the average curve generated by the arithmetic means of the individual occupant evacuation times for run 1 and run 2, M_3 represents the average curve generated by the arithmetic means of the individual occupant evacuation times for run 1, run 2 and run 3. M_4 represents the average curve generated by the arithmetic means of the individual occupant evacuation times for run 1, run 2, run 3 and run 4.

The consecutive ERD_j can be interpreted as a series convergent to the expected value equal to 0 (the case of two curves identical in magnitude). Hence, a measure of the convergence of the series is possible. A measure of the convergence of two consecutive Euclidean Relative Differences ERD_j, corresponding to two consecutive average curves $\overrightarrow{M_j}$ can be obtained by calculating ERD_{convj} (see Equation 20). It is expressed as the absolute value of the difference of two consecutive Euclidean Relative Differences, ERD_j and ERD_{j-1}.

$$ERD_{convj} = |ERD_j - ERD_{j-1}| \qquad \text{[Equation 20]}$$

The last ERD_{convj} value, corresponding to the differences between the latest average curves is $ERD_{convFIN}$ (see Equation 21).

$$ERD_{convFIN} = |ERD_p - ERD_{p-1}| \qquad \text{[Equation 21]}$$

Calculation of ERD_{convj} permits estimation of the impact of the number of runs on the overall differences between consecutive average curves. $ERD_{convFIN}$ represents therefore a tool to understand the uncertainty (e.g. *behavioural uncertainty*) associated with multiple runs of an individual evacuation scenario.

Convergence measure 4: Euclidean Projection Coefficient (EPC)
The same type of convergence measures can be produced for the Euclidean Projection Coefficient (EPC).

The consecutive EPC_j can be interpreted as a series convergent to the expected value equal to 1 (the best possible agreement between two consecutive EPC_j). Hence, a measure of the convergence of the series can be performed. This results in Equation 22 and 23.

$$EPC_{convj} = |EPC_j - EPC_{j-1}| \qquad \text{[Equation 22]}$$
$$EPC_{convFIN} = |EPC_p - EPC_{p-1}| \qquad \text{[Equation 23]}$$

EPC_{convj} permits the estimation of the impact of the number of runs on the possible agreement between two consecutive average curves. $EPC_{convFIN}$ is therefore another indicator of the *behavioural uncertainty* associated with multiple runs of an individual evacuation scenario.

Convergence measure 5: Secant Cosine (SC)

Convergence measures can be developed for the Secant Cosine (SC). The consecutive SC_j can be interpreted as a series convergent to the expected value equal to 1 (the case of two identical shapes of consecutive curves). Hence, a measure of the convergence of the series can be performed and it is presented in Equations 24 and 25.

$$SC_{convj} = |SC_j - SC_{j-1}|$$ [Equation 24]

$$SC_{convFIN} = |SC_p - SC_{p-1}|$$ [Equation 25]

SC_{convj} allows understanding of the impact of the number of runs on the possible differences between the shapes of two consecutive average curves. $SC_{convFIN}$ represents therefore a variable to understand the behavioural uncertainty associated with the average shape of the simulated curves, given a certain number of runs n of the same evacuation scenario. The shape of an occupant-evacuation curve enhances the understanding of the full evacuation process rather than the total evacuation time only.

4.1.3 The evaluation method

Five variables have been presented in the previous section, namely $TET_{convFIN}$, $SD_{convFIN}$, $ERD_{convFIN}$, $EPC_{convFIN}$, and $SC_{convFIN}$. These variables characterise the total evacuation times and the occupant-evacuation time curves. They are used to evaluate the convergence of the evacuation results towards average values. Those variables represent the basis for a novel evaluation method. The proposed method addresses two key aspects of evacuation modelling:

1) The analysis of behavioural uncertainty of a particular evacuation scenario.
2) The identification of the minimum number of runs to produce a stable evacuation curve of the same scenario in relation to the evacuation scenario and the model in use.

An iterative method is suggested for the evaluation of evacuation model results. The method is based on five steps (see Figure 18).

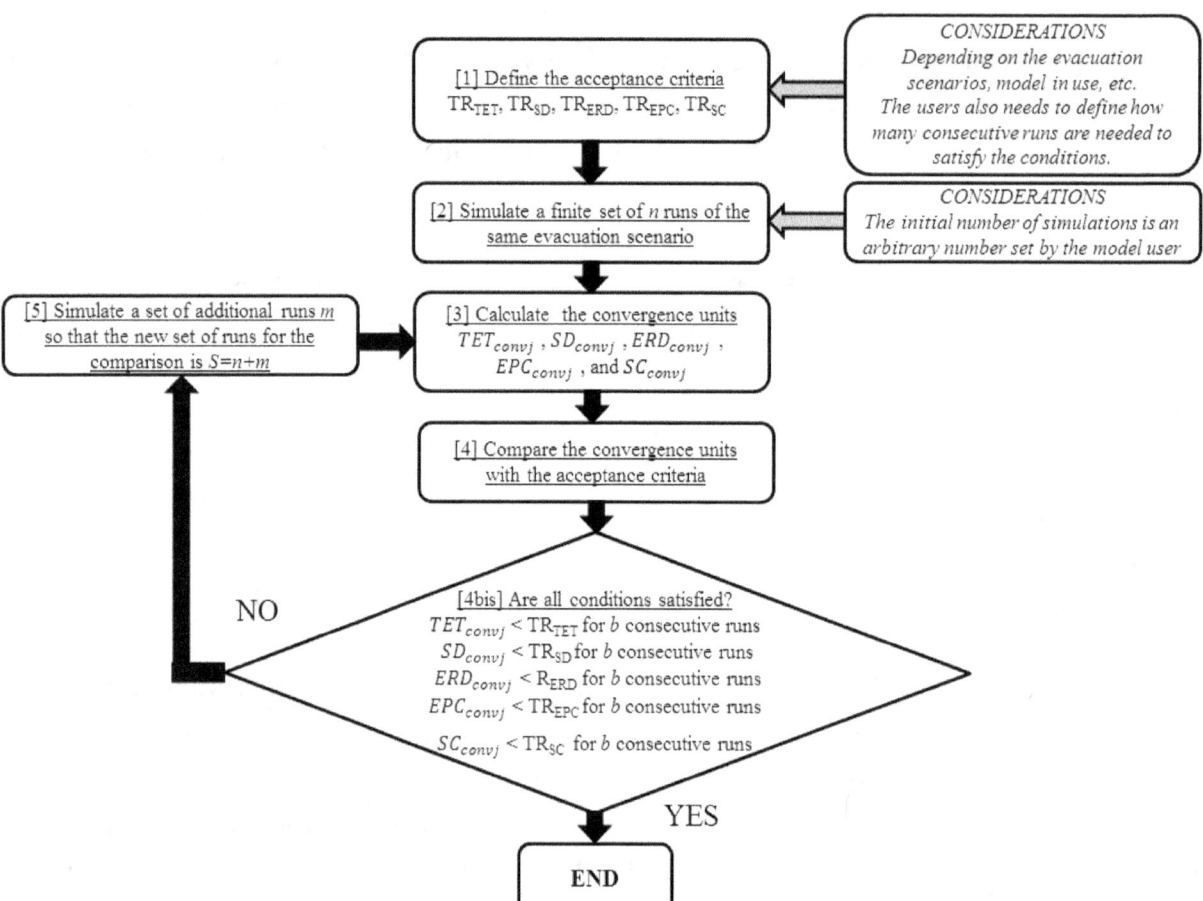

Figure 18. Schematic flow chart of the proposed evaluation method.

Step 1. <u>Define the acceptance criteria</u>. (see [1] in Figure 18)

The first step of the method consists of the identification of the acceptable thresholds to be achieved, i.e. the accepted behavioural uncertainty associated with the average curve obtained by multiple runs of the same scenario. This is associated with the user justification of the use of particular values for the distributions of behavioural options. In fact, the uncertainty associated with the selected input ought to impact the selection of the convergence criteria. The aim is to obtain an evacuation curve that is sufficiently stable given the scope of the analysis. For example, in the case of the use of evacuation modelling in the context of performance based design, the identification of these acceptable thresholds can be based on the estimated uncertainty during the calculation of the ASET (Available Safe Escape Time) produced using a fire model. This approach permits a joint analysis of the uncertainty associated with both the fire and evacuation simulations. Five thresholds (corresponding to the five convergence measures) are identified, namely TR_{TET}, TR_{SD}, TR_{ERD}, TR_{EPC}, TR_{SC}. It should be noted that there is an additional acceptance criteria that needs to be assessed, i.e., a finite number of consecutive runs b for which the acceptable thresholds must not be crossed. This needs to be assessed in order to verify that the convergence measures are stable under certain thresholds over a pre-defined number of runs. This requirement is based on the assumptions described in Section 4.1. A larger value for b would lead to a higher confidence that the acceptance criteria will be satisfied.

The identification of the acceptance criteria may depend on several factors such as the evacuation scenario, the model in use, uncertainty of input parameters, etc. The selection of the acceptance criteria - which may or may not include all convergence measures - may be identified by the evacuation modeller itself or from a third party.

Step 2. <u>Simulate a finite set of n runs of the same evacuation scenario</u> (see [2] in Figure 18)

Evacuation model users select an arbitrary initial number of simulations of an individual evacuation scenario, i.e., the same model input is used. n vectors $\vec{m}_{ij} = (m_{11}, \dots, m_{ij}, \dots, m_{qn})$ corresponding to the simulated evacuation times of each occupant i in each of the j runs are obtained. The occupant-evacuation time curves are produced, ranking the occupants in relation to their evacuation time.

The vector corresponding to the consecutive average curves $\vec{M} = (M_1, \dots, M_n)$ is also generated. In order to optimize the iterative process, the selection of the initial arbitrary number of runs may be based on a qualitative evaluation made by the evacuation modeller of the variability of the predicted outcome given the model input of the scenario under consideration (e.g. based on statistical considerations and sample size). Nevertheless, this judgment - which is the current qualitative method adopted by evacuation modellers to estimate the optimal number of runs - is not mandatory, since the proposed method permits a quantitative study of the impact of the number of runs on the occupant-evacuation time curve produced by the model.

Step 3. <u>Calculate the convergence measures</u> (see [3] in Figure 18)

The convergence measures presented in the previous sections are calculated for all runs, i.e., TET_{convj}, SD_{convj}, ERD_{convj}, EPC_{convj}, and SC_{convj}.

In order to perform the calculation of the secant cosines for all runs, model users need also to identify a finite set of values for s, needed for the calculation of SC_{convj}. As described in Section 4.1.1, the choice of the values for s relies on the dataset under consideration. SC_{convj} are calculated for all runs for as many s values as chosen by the model user.

Step 4-4bis. <u>Compare the convergence measures with the acceptance criteria</u> (see [4-4bis] in Figure 18)

The model user compares the calculated convergence measures against the acceptable thresholds defined during Step 1. This produces five tests that need to be accomplished:

TEST 1:

$$TET_{convj} < TR_{TET} \text{ for } b \text{ consecutive number of runs} \qquad \text{[Equation 26]}$$

TEST 2:

$$SD_{convj} < TR_{SD} \text{ for } b \text{ consecutive number of runs} \qquad \text{[Equation 27]}$$

TEST 3:

$$ERD_{convj} < TR_{ERD} \text{ for } b \text{ consecutive number of runs} \qquad \text{[Equation 28]}$$

TEST 4:

$$EPC_{convj} < TR_{EPC} \text{ for } b \text{ consecutive number of runs} \qquad \text{[Equation 29]}$$

TEST 5:

$$SC_{convj} < TR_{SC} \text{ for } b \text{ consecutive number of runs} \qquad \text{[Equation 30]}$$

It should be noted that the criteria need to be satisfied for a pre-defined finite number of consecutive b runs (as defined during Step 1). The values corresponding to the jth run where the conditions are verified for b consecutive runs represent $TET_{convFIN}$, $SD_{convFIN}$, $ERD_{convFIN}$, $EPC_{convFIN}$, and $SC_{convFIN}$.

If the five conditions are all satisfied for a pre-defined number of consecutive runs, the curves generated by n runs meet the acceptance criteria, i.e. the average curve is estimated given an accepted behavioural uncertainty associated with the number of runs (based on the acceptance criteria). If one or more of the conditions are not satisfied, the model user needs to proceed with Step 5.

Step 5. <u>Simulate a set of additional simulations m, so that the new set of runs for the comparison is $S=n+m$.</u> (see [5] in Figure 18)

The model user sets an arbitrary number of additional simulations to be run. The definition of the additional runs can be set in accordance with a qualitative analysis of any failed tests (see Equations 26-30). A new set of $S=n+m$ $\vec{S}_{ij}$ vectors $\vec{S}_{ij} = (S_{11}, ..., S_{ij}, ..., S_{q\,S})$ corresponding to the average simulated evacuation times of each occupant i in each of the j runs are obtained. The same methodology of Step 2 is adopted to produce the occupant-evacuation time curves, i.e., the occupants are ranked in relation to their evacuation time. The model user can now re-start the procedure starting from Step 3.

4.1.4 Example application of the evaluation method

An application of the method presented in the previous section is described to provide an example of the concepts. Given the explanatory scope of the example, data used in this section are fictitious, i.e., they do not correspond to real data. This choice has been driven by the current lack of repeated experimental data, i.e. the method has been applied to study simulation results. Data are created in order to be representative of the results obtained with an evacuation model for a hypothetical evacuation scenario. A fictitious set of numbers is produced using Wichman and Hill's [1982] pseudo-random generator. The pseudo-random numbers are used as input to produce lognormal-distributed values. This choice was made in order to be representative of a hypothetical evacuation scenario which is influenced by pre-evacuation times (which generally follow a log-normal distribution [Purser and Bensilum, 2001], The fictitious data are then used to create fictitious individual evacuation times calculated by progressively summing the values obtained (in order to be representative of a hypothetical real case study where evacuation ranges approximately between 1100 s and 1900 s). For example, if the first pseudo-random generated number is 12 s and the second pseudo-random generated number is 18 s, the evacuation time of the first occupant out would correspond to 12 s and the evacuation time of the second occupant out would be 12 s + 18 s = 30 s. The procedure is repeated for all 120 occupants (See Table 3). An example of one possible curve is provided in Figure 19. The assumed population consists of 120 occupants. The evaluation of the number of runs to be simulated is the unknown variable of this example.

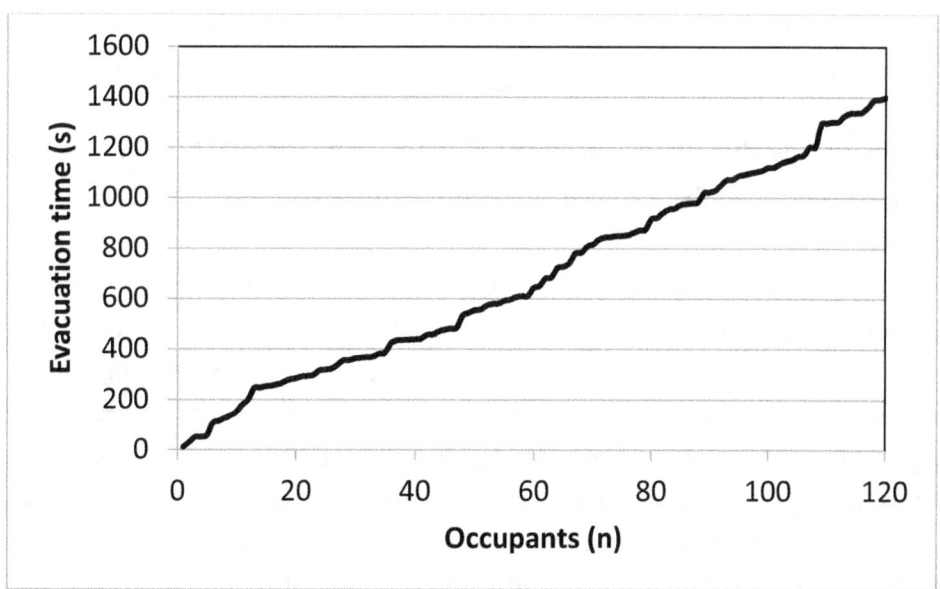

Figure 19. Fictitious data representing one possible curve of evacuation times.

Table 3. Example of fictitious data representing one possible occupant-evacuation curve.

Occupants out	Pseudo-random generated number	Evacuation time (s)
1	12	12
2	18	30
3	21	51
4	8	59
...	...	...
120	...	1401

The steps of the evaluation method are applied as follows.

Step 1. <u>Define the acceptance criteria.</u>

This step deals with the definition of the five acceptable thresholds TR_{TET}, TR_{SD}, TR_{ERD}, TR_{EPC}, TR_{SC} about the impact of the number of runs on the predicted outcome of the evacuation model for the same evacuation scenario (see Equations 31-35). The number of consecutive runs ($b=10$) for which the acceptance thresholds needs to be accomplished is also defined.

$$TR_{TET} = 0.5\% \qquad \text{[Equation 31]}$$
$$TR_{SD} = 5\% \qquad \text{[Equation 32]}$$
$$TR_{ERD} = 1\% \qquad \text{[Equation 33]}$$
$$TR_{EPC} = 1\% \qquad \text{[Equation 34]}$$
$$TR_{SC} = 1\% \qquad \text{[Equation 35]}$$

For instance, this means that the acceptance criteria are satisfied if $TET_{convj} < TR_{TET}$ for 10 consecutive runs, $SD_{convj} < TR_{SD}$ for 10 consecutive runs, etc.

It should be noted that the acceptance criteria have been selected with the only purpose of showing the procedure, i.e., they do not represent recommended values for use in real engineering analyses. Nevertheless, those criteria represent possible values in the context of fire safety engineering and all types of uncertainty associated with modelling results. In fact, the authors argue that thresholds below 5 % would permit the assessment of the required safe egress time with a reasonable degree of accuracy. The definition of the criteria would be dependent on several factors, such as the type of evacuation scenario, data under consideration, the scope of the analysis, the uncertainty in input parameters and their distribution, etc. In practice, modellers can check the convergence of the measures over the runs and calculate the progressive difference between the threshold value and the current values of each convergence measure. A modeller may also set a percentage of admitted difference between the thresholds in the b consecutive runs.

Step 2. <u>Run a finite set of n runs of the same evacuation scenario</u>

An arbitrary initial number of simulations of the same scenario is set to 35. *n=35* vectors of 120 dimensions $\vec{m}_{ij} = (m_{1\,1}, ..., m_{i\,j}, ..., m_{120\,35})$ corresponding to the simulated evacuation times of each ith occupant (for a total of 120 occupants) in each jth run are obtained (for a total of 35 runs).

In the present example, the 35 fictitious curves have been generated using the method described at the beginning of Section 3. They result in the 35 curves showed in Figure 20. The curves presented in Figure 20 are representative of a set of repeated results of an evacuation model in the case of a hypothetical evacuation scenario for lognormal distribution of evacuation times [Ronchi and Nilsson, 2013]. It should be noted that the shape of the evacuation curves may be different than the example provided here (e.g. s-shaped occupant-evacuation curves). The method is based on convergence measures which are independent on the shape of the curves, and thus may be applicable for any type of curve.

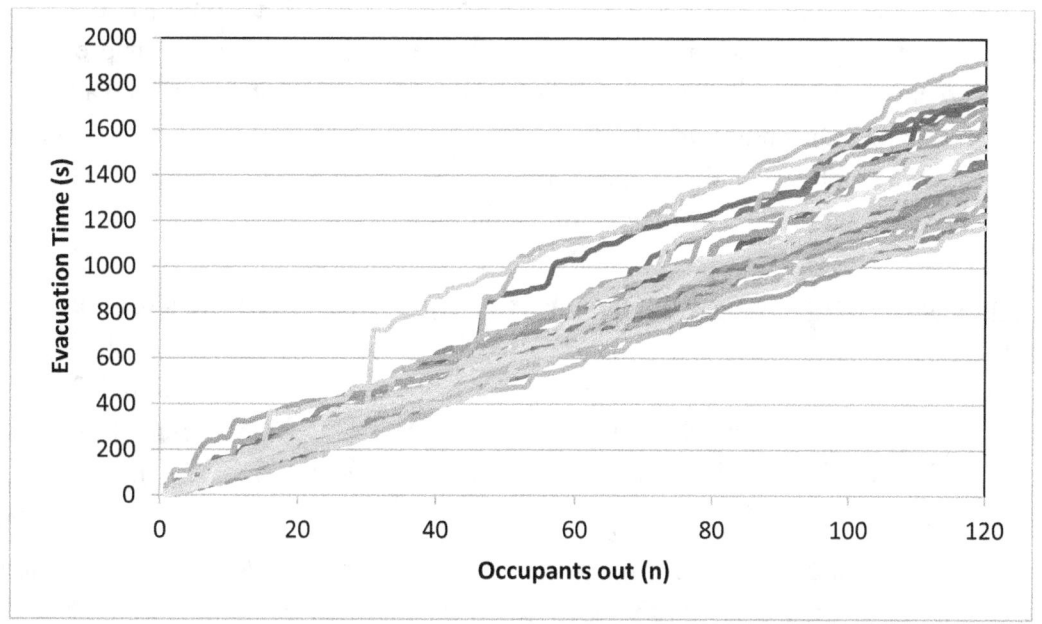

Figure 20. Fictitious data representing 35 runs of the same hypothetical evacuation scenario.

The vector corresponding to the consecutive average curves $\vec{M}=(M_1, ..., M_{35})$ is also generated.

61

Step 3. <u>Calculate the convergence measures</u>

The convergence measures presented in the previous sections are calculated for all 35 runs, i.e., TET_{convj}, SD_{convj}, ERD_{convj}, EPC_{convj}, and SC_{convj} in accordance to Equation 16, Equation 18, Equation 19, Equation 21, and Equation 24, respectively. In this example, a single value for s in Equation 24 has been used, namely $s=4$. Results are presented in Table A.1 in Appendix A.

Step 4-4bis. <u>Compare the convergence measures with the acceptance criteria</u>

Results for 35 runs are compared with the acceptance criteria defined in Step 1 (see also Equations 25-29). Table A.2 in Appendix A shows the results of the tests in relation to the number of runs. When the box shows "FAILED", it means that the test is failed. When the test is passed, the box is left blank. After 10 consecutive runs (given the acceptance criteria defined in Step 1), when the test is passed, the box shows "OK", which means that the acceptance criteria have been met.

In this example, Test 1 failed, Test 2 is passed after 25 runs, Test 3 is accomplished after 26 runs, Test 4 is failed, and Test 5 is accomplished after 15 runs. This means that our predicted curve meet the acceptance criteria with regards of the standard deviation of the total evacuation time, the Euclidean Relative Difference and the Secant Cosine. Nevertheless, there are two criteria that have not been met (Total Evacuation Time and Euclidean Projection Coefficient). It is, therefore, necessary to proceed with Step 5 by conducting additional runs.

Step 5. <u>Simulate a set of additional simulations m so that the new set of runs for the comparison is $S=n+m$.</u>
Another set of runs $m=35$ of the same scenario – corresponding to additional 35 occupant-evacuation time curves – are considered for a total of $S=n+m=35+35=70$ runs. In this example, additional fictitious data are produced using the same method as the first 35 curves. A new set of $S=n+m$ $\vec{S}_{ij}$ vectors $\vec{S}_{ij} = (S_{11}, \ldots, S_{ij}, \ldots, S_{q\,s})$ corresponding to the average simulated evacuation times of each of the 120 occupants i^h in each of the 70 j runs for which S are determined. The evaluation method is now repeated for $S=70$ runs, starting from Step 3, called here Step 3.2.

Step 3.2 <u>Calculate the convergence measures</u>
The failing convergence measures are calculated for $S=70$ runs, i.e., $TET_{convFIN}$, and $EPC_{convFIN}$ for our case study.

Step 4.2-4.2bis. <u>Compare the convergence measures with the acceptance criteria</u>

Results for $S=70$ runs are compared again with the acceptance criteria defined in Step 1. Table A.3 in Appendix A shows the results of the tests that were previously failing in relation to the number of runs.

Table A.3 shows that Test 4 is accomplished after 40 runs. An example of the number of runs required to accomplish different criteria for TR_{TET} (where $TET_{convj} < TR_{TET}$ for 10 consecutive runs) for the fictitious data-set under consideration is shown in Figure 21. The grey vertical line refers to the acceptance criteria $TR_{TET} = 0.5\%$ which has been selected for the analysis of the total evacuation time in Step 1. Test 1 is passed after 61 runs if the convergence criteria are $TET_{convj} < 0.5\%$ for 10 consecutive runs. This means that our predicted curve now meet all acceptance criteria.

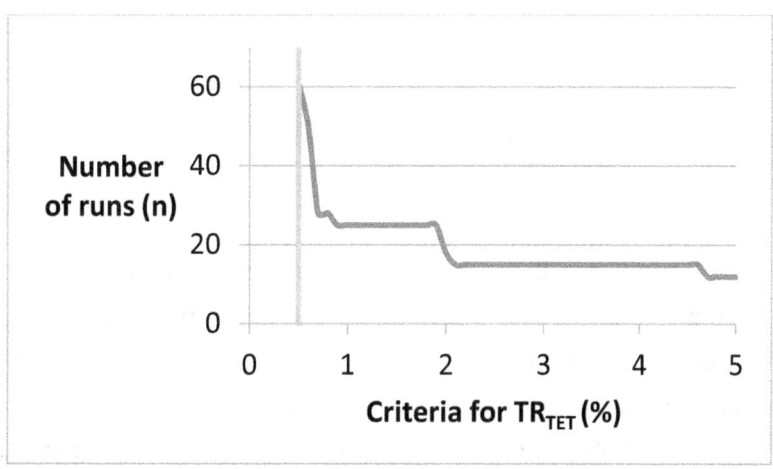

Figure 21. Number of required runs in relation to different criteria for TR_{TET}

The analysis of the trend of the convergence measures is useful to obtain general information on the type of data-set under consideration. For example, it is possible to assess behavioural uncertainty and therefore estimate the impact of the use of stochastic variables/distributions on evacuation model results.

An example from the data of the case study in Section 4.1.3 is presented in Figure 22 where TET_{convj} and SD_{convj} are shown as well as Figure 23 where ERD_{convj}, EPC_{convj}, and SC_{convj} are shown (convergence measures are calculated for a total of 140 consecutive average number of runs, i.e., 70 additional runs have been analysed).

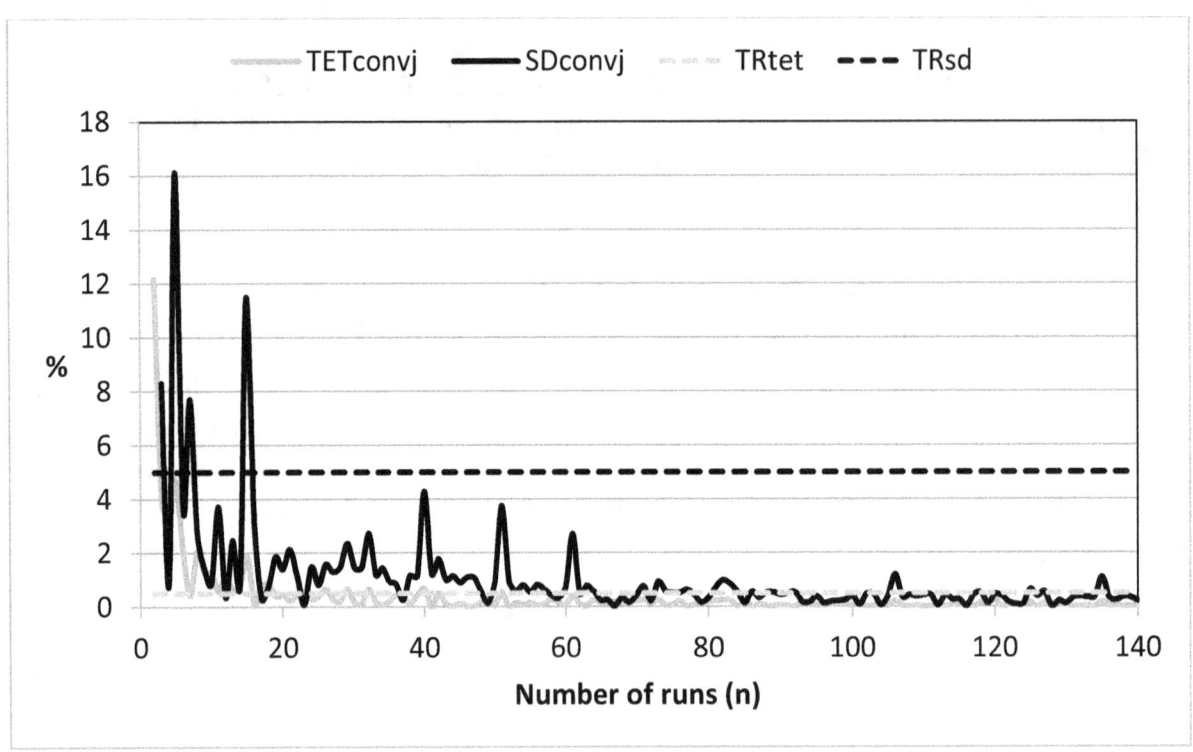

Figure 22. TET$_{convj}$, SD$_{convj}$ in relation to the consecutive average number of runs (expressed in %).

Figure 22 and Figure 23 show that the standard deviation of the evacuation time SD_{convj} of the last occupant is the slowest converging variable for the case study shown here. Together with TET_{convj}, those variables are useful to understand the variability of the total evacuation time in relation to the number of runs. An estimation of uncertainty (including behavioural uncertainty) associated with the total evacuation time is a key aspect of fire safety engineering analysis since it represents the RSET (Required Safe Egress Time) [Gwynne et al., 2012b], the time needed by all occupants to perform a safe evacuation.

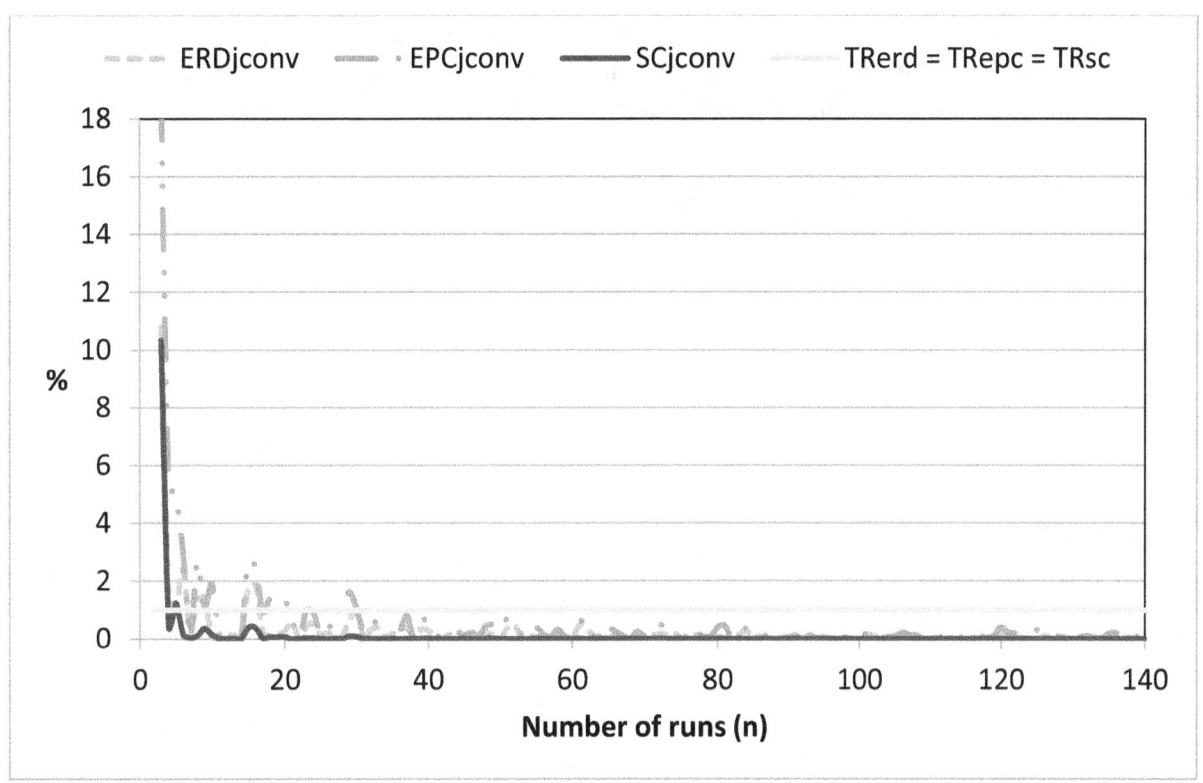

Figure 23. ERD_{convj} , EPC_{convj} , and SC_{convj} in relation to the consecutive average number of runs (expressed in %).

The analysis of the convergence of ERD_{convj} , EPC_{convj} , and SC_{convj} is also a significant contribution to the understanding of behavioural uncertainty, since it permits the analysis of the variability of the predicted occupant-evacuation time curves in relation to the number of runs. These measures permits the study of the entire occupant-evacuation curve rather than an analysis based only on the study of the total evacuation time. In the example provided here, the convergence measures are below 2.5 % after 17 runs, thus permitting the estimation of the average occupant-evacuation time curve with an admitted 2.5 % variability in a relatively small number of runs.

The simulation of an additional 70 runs (for a total of 140 runs in Figure 22 and Figure 23) shows that, as expected, results continue to converge and the effect of behavioural uncertainty on average occupant evacuation time is progressively reduced. Nevertheless, if the acceptance criteria include the requirement of being below the thresholds for a sufficient number of consecutive runs (i.e. a critical number that the model user should select in relation to the scenario under consideration in order to verify the stability of the convergence), the simulation of additional runs does not provide any additional benefits to the modeller. The selection of the number of runs is optimised in relation to the pre-defined acceptance criteria and there is no need to simulate additional runs.

A statistical estimation of the uncertainty associated with the use of the convergence measures can be performed in relation to the number of runs. This includes the study of the uncertainty of the sample average total evacuation times and the sample standard deviations.

Assume that each total evacuation time in the vector $\overrightarrow{TET}$ is a sum of random variables corresponding to the inter-temporal times between each occupant. Applying the central limit theorem, the series corresponding to the vector $\overrightarrow{TET}$ consists of assumed normally distributed values $TET_j \sim N(\mu, \sigma^2)$, where μ is the true mean value and σ^2 is the true variance. The sample variance is:

$$s^2 = \frac{\sum_{j=1}^{n}(TET_j - TET_{avj})^2}{n-1}$$ [Equation 36]

Where n is the number of runs. Applying Cochran's theorem, $s^2 \sim \frac{\sigma^2}{n-1} \chi^2_{n-1}$, which is a chi-squared distribution with n-1 degrees of freedom. Then, the variance of the sample variance, $Var(s^2)$, corresponds to:

$$Var(s^2) = Var\left(\frac{\sigma^2}{n-1}\chi^2_{n-1}\right) = \left(\frac{\sigma^2}{n-1}\right)^2 Var(\chi^2_{n-1}) =$$
$$= \left(\frac{\sigma^2}{n-1}\right)^2 2(n-1) = \frac{2\sigma^4}{n-1}$$ [Equation 37]

The sample standard deviation s is distributed as a chi distribution with n-1 degrees of freedom, i.e., $s \sim \frac{\sigma}{\sqrt{n-1}}\chi_{n-1}$. Hence the variance of the standard deviations of the sample data corresponds to:

$$Var(s) = Var\left(\frac{\sigma}{\sqrt{n-1}}\chi_{n-1}\right) = \frac{\sigma^2}{n-1} Var(\chi_{n-1}) =$$
$$= \frac{\sigma^2}{n-1}\left[n - 1 - 2\left\{\frac{\Gamma\left(\frac{n}{2}\right)}{\Gamma\left(\frac{n-1}{2}\right)}\right\}^2\right]$$ [Equation 38]

Where $\Gamma(n)$ is a gamma function. Hence, it is possible to estimate the relative standard deviation (the relative difference between the use of sample standard deviations and the standard deviations corresponding to the true distribution):

$$relative\ Std(s) = \sqrt{\frac{\left[n-1-2\left\{\frac{\Gamma\left(\frac{n}{2}\right)}{\Gamma\left(\frac{n-1}{2}\right)}\right\}^2\right]}{n-1}}$$ [Equation 39]

This information permits an estimation of the uncertainty associated with the use of the estimate standard deviations SD_j employed in the evaluation method in relation to the number of runs under consideration.

It is also possible to perform an estimation of the uncertainty associated with the use of the estimate variance of the sample data s^2 when calculating the average sample Total Evacuation

66

Time TET_{avj}. In fact, the average of the Total Evacuation Time TET_{avj} is distributed as $\frac{s}{\sqrt{n}}t_{n-1} + \mu$, where t_{n-1} is a Student t random variable with n-1 degrees of freedom.

Therefore the variance of the sample average TET_{avj} corresponds to:

$$Var\left(TET_{avj}\right) = \frac{s^2}{n}\frac{n-1}{n-3} \qquad \text{[Equation 40]}$$

And the uncertainty of the TET_{avj} is:

$$SD\left(TET_{avj}\right) = s\sqrt{\frac{n-1}{n(n-3)}} \qquad \text{[Equation 41]}$$

It is therefore possible to estimate the uncertainty associated with the number of runs given the use of the sample TET_{avj}.

To date, behavioural uncertainty is generally treated only in a qualitative manner (performing a qualitative evaluation of the number of runs to be simulated). It is argued that the present work would encourage evacuation model users to perform a quantitative treatment of this type of uncertainty given the simplicity of the method proposed. The proposed method may appear costly if compared with the simulation of a low arbitrary number of runs. Nevertheless, it permits saving computational time since it allows the estimation of the exact number of runs needed to obtain a pre-defined level of accuracy.

The benefits obtained from the use of the method apply to design studies as well as model validation. The proposed method permits an estimation of the convergence of the simulated occupant-evacuation curve towards the average curve, thus increasing the significance of the model predictions. This is reflected in a better understanding of the variability of RSET and the possible estimation of the margin of safety of a specific design in relation to behavioural uncertainty.

From a model validation perspective, to date, two antithetical approaches may be used to present model comparison with experimental data, namely 1) the use of the best model estimation for the occupant-evacuation time curve, or 2) the average occupant-evacuation time curve. The method presented in this section increases the usability of the second approach, since it allows a thorough quantitative understanding of the average curves produced by evacuation models. Future work based on the presented method is therefore a definition of an evacuation model validation protocol which uses the convergence measures to assess the differences between model predictions and experimental data by taking into account behavioural uncertainty.

A possible additional application of the method presented in this document may be its use for the comparison of model predictions produced by different evacuation models. It would be in fact possible to quantify the impact of the stochastic variables and assumptions used by different evacuation models given the same evacuation scenario. In addition, the same approach can be used to evaluate other type of evacuation model results such as congestion levels, travel distance, etc.

A set of limitations of the proposed method can be identified both in terms of its assumptions as well as its applicability.

The first limitation of the method is that it uses the concepts of convergence in mean and the central limit theorem rather than a statistical estimation of the expected values. Hence, the choice of the requirement for the finite number of consecutive runs b for which acceptable thresholds must not be crossed should be carefully evaluated by the modeller in relation to the data under consideration. This limitation is tempered by the simplicity of the proposed method, i.e., it can be applied by evacuation modellers to analyse behavioural uncertainty without a complex inferential statistical treatment of the data, which may require time and user expertise.

Another limitation of the method is associated with the assumptions that evacuation curves can be identical between model runs even in the case of different behaviours, i.e., the arrival rates to the exits are the same but they refer to different occupants or different exits.

With regards of the method applicability, multiple data-sets of a single evacuation scenario are rarely available in the literature. This makes it difficult to study the impact of behavioural uncertainty on experimental data. Given the current stage of experimental evacuation research, the proposed method is mainly applicable to the analysis of behavioural uncertainty in simulation results. Once additional experimental data on individual scenarios will be available, researchers will be able to use the same concepts introduced in this paper for the analysis of behavioural uncertainty in experimental data.

Without multiple experimental data, a single experiment often represents the only reference on that specific evacuation scenario, but it is not clear whether it represents average behaviour or it is an outlier. In fact, the assessment of experimental and evacuation model results may also include the analysis of the tails of the distribution rather than the analysis of the peaks (i.e. average values). Nevertheless, the authors argue that the study of the average model predictions together with the variability of results around the average is deemed to be a useful method to analyse *behavioural uncertainty*. The human behaviour in fire research community is aware of the lack of experimental data and the need to fill this gap with data collection efforts [20]. In recent years, significant data collection efforts have been carried out (e.g. several projects were performed for different aspects/conditions of the evacuation process using several tool to aid the collection and quality of data [Gwynne, 2013]). Therefore, it can be argued that considering a long-term perspective, it will be possible to assess behavioural uncertainty also for experimental data (thus making the method proposed in the paper applicable also for that issue).

The method is presented using a case study based on pseudo-random generated numbers. Future work can be based on the analysis of the results of an evacuation model from a real world case study. For instance, if repeated experimental data are available (which will allow expanding the list of data-sets in Section 3.2.1), the method will be useful to evaluate behavioural uncertainty during the performance of validation tests. Therefore, the need for the collection of repeated experimental data is deemed to be a necessary step in order to perform reliable validation studies.

5.0 Discussion on the Verification and Validation Protocol

This section discusses the definition of a standard V&V protocol for evacuation modelling. The scope of this section is not to provide acceptance criteria about model results, but to open a discussion on the issues associated with their definition.

To date, there is no standard V&V protocol for the assessment of building evacuation model results. Little research has been conducted on the methods for the evaluation of the predictive capabilities of building evacuation models. Lord et al. [2005] discussed the issues associated with uncertainty and variability in egress data and computational methods for egress analysis. In particular, they apply and refine a method of uncertainty analysis. This method includes sensitivity analysis for the study of the impact of different model parameters on the results generated by the models (i.e., *model input uncertainty* [Hamins and McGrattan, 2007]). Evacuation times represent the main variable under consideration in Lord et al. [2005]. Galea et al. [2012a] pointed out that simply predicting evacuation time for the overall population is not sufficient to determine the accuracy of the representation of the evacuation process. For this reason, they applied the concept of functional analysis presented by Peacock et al. [1999] and in the ISO document 16730 [International Standards Organization, 2008] for the comparison between model predictions and experimental data-sets for the complete occupant-evacuation time curves. Galea et al.'s [2012a] method permits the assessment of the predictive capabilities of evacuation models analysing the full evacuation process. Section 4 of this document presents a method based on convergence criteria for the study of behavioural uncertainty. This method addresses the study of behavioural uncertainty, i.e., the variability of evacuation model predictions in relation to the number of runs and the use of stochastic algorithms/distributions. It should also be noted that the use of the proposed method is not dependent on the type of algorithms/distributions employed by the model under consideration, i.e. the method can be applied for any evacuation model.

Two antithetical approaches may be used to compare model results with experimental data, namely 1) the use of the best model estimation for the occupant-evacuation time curve [Galea et al., 2012b], or 2) the average occupant-evacuation time curve. The method presented in Section 4 of this document increases the usability of the second approach, since it allows a thorough quantitative understanding of the average curves produced by evacuation models. It permits an estimation of the convergence of the simulated occupant-evacuation curve towards the average curve, thus increasing the understanding of the model predictions. This is reflected in a better understanding of the variability of RSET and the possible estimation of the margin of safety of a specific design in relation to behavioural uncertainty.

The current status of experimental evacuation research does not permit a full understanding of all uncertainties associated with experimental data on occupant behaviours during building fire evacuation (*measurement uncertainty, intrinsic uncertainty* and *behavioural uncertainty*). The impact of behavioural uncertainty on evacuation results may be scenario-dependent, but its assessment is crucial for a complete understanding of the evacuation process. Nevertheless, multiple data-sets of a single evacuation scenario are rarely available in the literature, thus making it difficult to assess behavioural uncertainty experimentally.

If repeated experimental data of single evacuations scenarios are available, it would be possible to compare multiple runs of a model to multiple curves of data. Therefore, different models may be validated by studying the convergence of their results towards the average in relation to the convergence of different experimental curves. The method presented here would permit the study of model predictions in relation to the agreement between the convergence of simulation results and the experimental data-sets.

The method presented in this document can also be used to compare different model predictions against each other. The method can be employed to rank different models in terms of their agreement with experimental data-sets, thus permitting the assessment of high performing models in relation to models with small values of the convergence measures.

The need for additional experimental data on building evacuation scenarios is evident in order to fully assess the acceptance criteria included in a Verification and Validation protocol. Several questions affect the definition of those acceptance criteria. How should acceptance criteria be defined in relation to the intended use of the model? How do we define acceptance criteria in relation to the current lack of knowledge on human behaviour in fire and uncertainty associated with human factors data-sets? Who should set the criteria? (A) model developers, (B) a third party (e.g. Institutional organizations, e.g. International Maritime Organization, International Standards Organization, etc.), (C) the model users, or (D) a joint effort of all the parties involved? These questions do not have simple answers and they require a discussion between all parties involved.

The criteria will be dependent on two main factors, namely 1) the intended use of the model and 2) the uncertainty associated with the benchmark data and its nature. The second aspect is associated with the type of test under consideration, whether it is an ideal case, an experimental data-set or an actual evacuation data-set. Suggested acceptance criteria are already available in the literature for different contexts of use [Meyer-Koenig et al., 2007, Galea et al., 2012a]. Nevertheless, there is a need for a broad debate within the evacuation modelling community on whether they should be only minimum criteria (i.e. should the models accomplish the thresholds proposed, they will not automatically become "certified models" and it is a responsibility of the end users to evaluate the confidence to put into model predictions) and to which degree different parties should be involved in the criteria definition and V&V assessment.

References

Averill, J.D. (2011). Five Grand Challenges in Pedestrian and Evacuation Dynamics, in: Peacock, R.D., Kuligowski, E.D., Averill, Jason D. (Eds.), Pedestrian and Evacuation Dynamics. Springer US, Boston, MA, pp. 1–11.

Averill, J.D., Reneke, P.A., Peacock, R.D., (2008). Required Safe Egress Time: Data and Modeling, in: Sky Is the Limit. Society of Fire Protection Engineers, Bethesda, MD, Auckland, New Zealand, pp. 301–311.

Bayer, K. and Rejnö, T. (1999). Optimering genom fullskaleförsök [Evacuation alarm – Optimizing through full-scale experiments] (No. 5053). Department of Fire Safety Engineering, Lund University, Lund, Sweden.

Boyce, K.E., Purser, D.A., Shields, T.J. (2012). Experimental studies to investigate merging behaviour in a staircase. Fire and Materials 36, 383–398.

Chooramun, N. (2011). Implementing a hybrid spatial discretisation within an agent based evacuation model. University of Greenwich, London, UK.

Deutsch, M., Gerard, H.B. (1955). A study of normative and informational social influences upon individual judgment. The Journal of Abnormal and Social Psychology 51, 629–636.

Fang, Z.-M., Song, W.-G., Li, Z.-J., Tian, W., Lv, W., Ma, J., Xiao, X. (2012). Experimental study on evacuation process in a stairwell of a high-rise building. Building and Environment 47, 316–321.

Frantzich, H. and Nilsson, D., (2003). Utrymning genom tät rök: beteende och förflyttning [Evacuation in dense smoke: behaviour and movement] Technical Report 3126.

Frantzich, H., Nilsson, D., Eriksson, O. (2007). Utvärdering och validering av utrymningsprogram [Evaluation and validation of evacuation programs] (No. 3143). Department of Fire Safety Engineering and Systems Safety, Lund University, Lund, Sweden.

Galea, E.R. (1997). University of Greenwich, Centre for Numerical Modelling and Process Analysis, Validation of evacuation models. CMS Press.

Galea, E.R., Deere, S., Brown, R., Nicholls, I., Hifi, Y., Bresnard, N. (2012a). The SAFEGUARD validation data-set and recommendations to IMO to update MSC Circ 1238. Presented at the SAFEGUARD Passenger Evacuation Seminar, The Royal Institution of Naval Architects, London, UK, pp. 41–60.

Galea, E.R., Deere, S., Brown, R., Filippidis, L. (2012b). An Evacuation Validation Data Set for Large Passenger Ships. Presented at the Pedestrian and Evacuation Dynamics 2012 Conference, ETH, Zurich (Switzerland).

Grosshandler, W.L., Bryner, N.P., Madrzykowski, D., Kuntz, K. (2005). Report of the Technical Investigation of The Station Nightclub Fire (NIST NCSTAR 2: Volume 1). National Institute of Standards and Technology, Gaithersburg, MD (US).

Gwynne, S.M.V., Kuligowski, E.D., Spearpoint, M. (2012a). More thoughts on defaults. Presented at the Fifth International Symposium on Human Behaviour in Fire, Interscience Communications, Cambridge, UK, pp. 9–23.

Gwynne, S., Kuligowski, E., Nilsson, D. (2012b). Representing evacuation behavior in engineering terms. Journal of Fire Protection Engineering 22, 133–150.

Gwynne, S.M.V. (2013). Improving the Collection and Use of Human Egress Data. Fire Technology 49, 83–99.

Hamins, A. and McGrattan, K. (2007). Verification and Validation of Selected Fire Models for Nuclear Power Plant Applications, Volume 2 (NUREG-1824). National Institute of Standards and Technology, Gaithersburg, MD (US).

Hewstone, M. and Martin, R. (2008). Social Influence, in: Introduction to Social Psychology. Blackwell Publishing, London, UK.

Hokugo, A., Kubo, K., Murozaki, Y. (1985). An Experimental Study on Confluence of Two Foot Traffic Flows in Staircase. Journal of Architecture, planning and Environmental Engineering 358, 37–43.

Hoogendoorn, S.P. and Daamen, W. (2005). Pedestrian Behavior at Bottlenecks. Transportation Science 39, 147–159.

International Code Council (2012). International Building Code 2012.

International Maritime Organization (2007). Guidelines for Evacuation Analyses for New and Existing Passenger Ships, MSC/Circ.1238.

International Standards Organization (2008). Fire Safety Engineering – Assessment, verification and validation of calculation methods. ISO 16730.

Jin, T. (2008). Visibility and Human Behavior in Fire Smoke, in: SFPE Handbook of Fire Protection Engineering (3rd Edition). National Fire Protection Association, Quincy, MA (USA), pp. 2–42 – 2–53.

Jones, W.W., Peacock, R.D., Forney, G., Reneke, P.A. (2009). CFAST – Consolidated Model of Fire Growth and Smoke Transport (Version 6). Technical Reference Guide. NIST Special Publication 1026.

Jones, W. W.; Peacock, R. D.; Reneke, P. A.; Forney, G. P.; Kostreva, M. M. (1995) Understanding Sensitivity Analysis for Complex Fire Models. Session II; Fire Dynamics 3; European

Symposium on Fire Safety Science, First (1st). Proceedings. Session II. Fire Dynamics 3. August 21-23, 1995, Zurich, Switzerland, II-24/113-114 pp.

Kinateder, M. (2013). Social Influence in Emergency Situations–Studies in Virtual Reality. Phd Dissertation.

Korhonen, T. and Hostikka, S. (2009). Fire Dynamics Simulator with Evacuation: FDS+Evac Technical Reference and User's Guide (Working paper No. 119). VTT Technical Research Center of Finland.

Kuligowski, E. (2011). Predicting Human Behavior During Fires. Fire Technology 49, 101–120.

Kuligowski, E.D., Peacock, R.D., Hoskins, B.L. (2010). A Review of Building Evacuation Models, 2nd Edition, NIST Technical Note 1680.

Kuligowski, E.D. and Peacock, R.D. (2010). Building Occupant Egress Data, Report of Test FR 4024. National Institute of Standards and Technology, Gaithersburg, MD (US).

Latané, B., Darley, J.M. (1970). The unresponsive bystander: why doesn't he help? Appleton-Century Crofts, New York.

Lord, J., Meacham, B., Moore, A., Fahy, R., Proulx, G. (2005). Guide for evaluating the predictive capabilities of computer egress models NIST GCR 06-886.

McGrattan, K., Hostikka, S., Floyd, J.E., Baum, H.R., Rehm, R.G. (2007). Fire Dynamics Simulator (Version 5): Technical Reference Guide (NIST Special Publication 1018-5 No. SP 1018). National Institute of Standards and Technology, Gaithersburg, MD (US).

Meyer-König, T., Waldau, N., Klüpfel, H. (2007). The RiMEA Project — Development of a new Regulation, in: Waldau, Nathalie, Gattermann, P., Knoflacher, H., Schreckenberg, M. (Eds.), Pedestrian and Evacuation Dynamics 2005. Springer Berlin Heidelberg, Berlin, Heidelberg, pp. 309–313.

Melly, M., Lennon, P., Lennon, R. (2009). Who defers to whom? Deference behaviour on stairs. Presented at the Human Behaviour in Fire 2009 Symposium, Interscience Communications, pp. 135–146.

Mulholland G.W. (2008) Smoke Production and Properties. In the SFPE Handbook of Fire Protection Engineering (Fourth Edition). National Fire Protection Association, Quincy MA, USA.

Nilsson, D.and Johansson, A. (2009). Social influence during the initial phase of a fire evacuation— Analysis of evacuation experiments in a cinema theatre. Fire Safety Journal 44, 71–79.

Nilsson, J. and Petersson, R. (2008). Utvärdering av videoanalysmetoder för utrymning med tillämpning på horn [Evaluation of video analysis techniques for emergency scenarios] (Technical Report No. 5256). Department of Fire Safety Engineering and Systems Safety, Lund University., Lund, Sweden.

Nilsson, D. (2009). Exit choice in fire emergencies: influencing choice of exit with flashing lights. Dept. of Fire Safety Engineering and Systems Safety, Lund University, Lund, Sweden.

Peacock, R.D., Hoskins, B.L., Kuligowski, E.D. (2012). Overall and local movement speeds during fire drill evacuations in buildings up to 31 stories. Safety Science 50, 1655–1664.

Peacock, R.D., Reneke, P.A., D. Davis, W., Jones, W.W. (1999). Quantifying fire model evaluation using functional analysis. Fire Safety Journal 33, 167–184.

Purser, D. and Bensilum, M. (2001). Quantification of behaviour for engineering design standards and escape time calculations. Safety Science 38, 157–182.

Purser, D.A. (2008). Assessment of Hazards to Occupants from smoke, toxic gases and heat, in: SFPE Handbook of Fire Protection Engineering (4th Edition). Di Nenno P. J., Quincy, MA (USA), pp. 2–96 – 2–193.

Ronchi, E. and Kinsey, M. (2011). Evacuation models of the future: Insights from an online survey on user's experiences and needs. Presented at the Advanced Research Workshop Evacuation and Human Behaviour in Emergency Situations EVAC11, Capote, J. et al, Santander, Spain, pp. 145–155.

Ronchi, E., Gwynne, S.M.V., Purser, D.A., Colonna, P. (2013a). Representation of the Impact of Smoke on Agent Walking Speeds in Evacuation Models. Fire Technol 49, 411–431.

Ronchi, E., Nilsson, D., Zechlin, O., Klein, W., Mayer, H. (2013b). Employing validation and verification tests as an integral part of evacuation model development. Presented at the 13th International Conference and Exhibition on Fire Science and Engineering, Interscience Communications, Royal Holloway College, University of London, UK, pp. 979–990.

Ronchi, E., Reneke, P.A., Peacock, R.D. (2013c). A Method for the Analysis of Behavioural Uncertainty in Evacuation Modelling. Fire Technol. Doi: 10.1007/s10694-013-0352-7

Ronchi, E. and Nilsson, D. (2013). Modelling total evacuation strategies for high-rise buildings. Build. Simul. Doi: 10.1007/s12273-013-0132-9

Rykiel, E.J. (1996). Testing ecological models: the meaning of validation. Ecological modelling 90, 229–244.

Seyfried, A., Rupprecht, T., Winkens, A., Passon, O., Steffen, B., Klingsch, W.W.F., Boltes, M. (2007). Capacity Estimation for Emergency Exits and Bottlenecks. Presented at the 11th International Fire Science and Engineering Conference Interflam 2007, Royal Holloway College, University of London, UK.

Sime, J.D. (1985). Movement toward the Familiar: Person and Place Affiliation in a Fire Entrapment Setting. Environment and Behavior 17, 697–724.

Weidmann, U. (1992). Transporttechnik der Fussgänger. Transporttechnische Eigenschaften des Fussgängerverkehrs Literaturauswertung. IVT, Institut für Verkehrsplanung, Transporttechnik, Strassen- und Eisenbahnbau, Zurich, Switzerland.

Zhang, J., Klingsch, W., Schadschneider, A., Seyfried, A. (2011). Transitions in pedestrian fundamental diagrams of straight corridors and T-junctions. Journal of Statistical Mechanics: Theory and Experiment 2011, P06004.

Appendix A. Results of the example application of the evaluation method for behavioural uncertainty

Table A.1. Results corresponding to 35 runs of the same evacuation scenario (expressed in %).

Run (n)	TETconvj (%)	SDconvj (%)	ERDjconv (%)	EPCjconv (%)	SCjconv (%)
1	/	/	/	/	/
2	12.163	/	/	/	/
3	3.852	8.294	10.800	18.674	10.337
4	3.369	0.886	1.555	5.446	0.439
5	4.691	16.110	1.059	4.962	1.238
6	2.024	3.639	2.399	3.022	0.176
7	0.405	7.698	0.368	0.052	0.045
8	2.054	2.709	1.321	2.658	0.115
9	1.323	1.400	1.403	1.086	0.365
10	1.238	0.875	0.055	1.900	0.165
11	0.582	3.723	0.158	0.088	0.024
12	1.139	0.358	0.169	0.187	0.011
13	0.626	2.473	0.122	0.138	0.008
14	0.820	0.772	0.081	0.068	0.004
15	1.901	11.480	1.761	2.673	0.392
16	0.140	3.194	1.883	2.505	0.396
17	0.758	0.312	0.235	0.969	0.042
18	0.861	0.779	0.134	1.393	0.064
19	0.403	1.859	0.166	1.248	0.062
20	0.453	1.400	0.254	1.355	0.062
21	0.234	2.124	0.578	0.696	0.003
22	0.420	1.159	0.028	0.092	0.010
23	0.569	0.080	0.443	0.980	0.032
24	0.298	1.473	0.514	0.893	0.029
25	0.393	0.819	0.170	0.206	0.012
26	0.663	1.576	0.206	0.210	0.002
27	0.267	1.297	0.138	1.033	0.016
28	0.198	1.480	0.084	0.955	0.006
29	0.666	2.348	0.666	1.631	0.095
30	0.166	1.442	0.958	1.043	0.079
31	0.129	1.481	0.031	0.231	0.020
32	0.652	2.721	0.323	0.359	0.012
33	0.184	1.193	0.382	0.646	0.013
34	0.075	1.441	0.044	0.075	0.001
35	0.207	0.956	0.181	0.424	0.012

Table A.2 Summary of the results of the tests in Step 4 of the evaluation method.

Run	TEST 1	TEST 2	TEST 3	TEST 4	TEST 5
1	/	/	/	/	/
2	FAILED	FAILED	/	/	/
3	FAILED	FAILED	FAILED	FAILED	FAILED
4	FAILED		FAILED	FAILED	
5	FAILED	FAILED	FAILED	FAILED	FAILED
6	FAILED		FAILED	FAILED	
7		FAILED			
8	FAILED		FAILED	FAILED	
9	FAILED		FAILED	FAILED	
10	FAILED			FAILED	
11	FAILED				
12	FAILED				
13	FAILED				
14	FAILED				
15	FAILED	FAILED	FAILED	FAILED	**OK**
16			FAILED	FAILED	
17	FAILED				
18	FAILED			FAILED	
19				FAILED	
20				FAILED	
21					
22					
23	FAILED				
24					
25		**OK**			
26	FAILED		**OK**		
27				FAILED	
28					
29	FAILED			FAILED	
30				FAILED	
31					
32	FAILED				
33					
34					
35					

Table A.3. Results of Test 1 and Test 4 for 70 runs.

Run	TEST 1	TEST 4	Run	TEST 1	TEST 4	Run	TEST 1	TEST 4
1			24			47		
2	FAILED		25			48		
3	FAILED	FAILED	26	FAILED		49		
4	FAILED	FAILED	27		FAILED	50		
5	FAILED	FAILED	28			51	FAILED	
6	FAILED	FAILED	29	FAILED	FAILED	52		
7			30		FAILED	53		
8	FAILED	FAILED	31			54		
9	FAILED	FAILED	32	FAILED		55		
10	FAILED	FAILED	33			56		
11	FAILED		34			57		
12	FAILED		35			58		
13	FAILED		36			59		
14	FAILED		37			60		
15	FAILED	FAILED	38			61	OK	
16		FAILED	39			62		
17	FAILED		40	FAILED	OK	63		
18	FAILED	FAILED	41			64		
19		FAILED	42	FAILED		65		
20		FAILED	43			66		
21			44			67		
22			45			68		
23	FAILED		46			69		
						70		

www.ingramcontent.com/pod-product-compliance
Lightning Source LLC
Chambersburg PA
CBHW081835170526
45167CB00007B/2821